Promoting Investment

IN AGRICULTURE FOR INCREASED PRODUCTION AND PRODUCTIVITY

Saifullah Syed and Masahiro Miyazako

Prepared under the Japan Trust Fund project Support to Study on Appropriate Policy Measures to Increase Investment in Agriculture and to Stimulate Food Production.

March 2013
Rome, Italy

CABI is a trading name of CAB International

CABI
Nosworthy Way
Wallingford
Oxfordshire OX10 8DE
UK

Tel: +44 (0)1491 832111
Fax: +44 (0)1491 833508
E-mail: info@cabi.org
Website: www.cabi.org

CABI
38 Chauncey Street
Suite 1002
Boston, MA 02111
USA

T: +1 800 552 3083 (toll free)
T: +1 (0)617 395 4051
E-mail: cabi-nao@cabi.org

© FAO, 2013.

A catalogue record for this book is available from the British Library, London, UK.

Library of Congress Cataloging-in-Publication Data

Syed, Saifullah.
  Promoting investment in agriculture for increased production and productivity / Saifullah Syed and Masahiro Miyazako.
    p. cm.
  Includes bibliographical references and index.
  ISBN 978-1-78064-388-5 (alk. paper) -- ISBN 978-9251075883 (FAO) 1. Agriculture--Developing countries--Finance. 2. Agriculture--Economic aspects--Developing countries. I. Miyazako, Masahiro. II. Title.

  HD1417.S94 2013
  338.1'3091724--dc23

          2013025683

Published jointly by CAB International and FAO
Food and Agriculture Organization of the United Nations (FAO)
Viale delle Terme di Caracalla, 00153 Rome, Italy
website: www.fao.org

ISBN: 978 1 78064 388 5 (CABI)
ISBN: 978 9 25107 588 3 (print) (FAO)
E-ISBN: 978 9 25107 589 0 (PDF) (FAO)

Graphic design and typesetting by Maxtudio, Roma/New York
Printed and bound by Gutenberg Press Ltd, Tarxien, Malta.

Contents

Foreword

Investing in agriculture is one of the most effective ways of reducing hunger and poverty, promoting agricultural productivity and enhancing environmental sustainability. However, for any investment to have a positive impact on agricultural production and productivity, it must contribute to capital formation at the farm level. In this respect, investments made by the farmers themselves are indispensable. Their investments constitute the foundation and the engine for sustainable development and the reduction of poverty and hunger.

For farmers, the main sources of investment finance are their own savings and their fixed capital, which are used as collateral for credit. Capital formation is certainly higher for farming households with positive savings and clear, legally recognized ownership of their land. In areas where the levels of poverty and hunger are high and agriculture is dominated by small-scale farmers, such as in South Asia, sub-Saharan Africa and parts of Latin America, the average farmer earns less than half of what is needed to cross the poverty line. For small and marginal farmers with below average land holdings, the situation is even worse, both in terms of their ability to save and to secure their rights to the land.

Apart from the capacity to invest through the generation of savings and fixed assets, the factors driving investment for farm-level capital formation are the growth of the food value chain from producers to consumers, which includes agro-industries and the provision of public goods in the form of basic infrastructure, such as roads, electricity, education and technology. There is no doubt that more public resources are needed for agriculture. However, there is a need for new investment strategies that are centred on agricultural producers and focuses public resources at all levels on the provision of public goods in ways that complement investments made by farmers and support inclusive and efficient agricultural and food systems at local and national level.

To address these issues, in October 2009 FAO initiated the project: Study on Appropriate Policy Measures to Increase Investments in Agriculture and to Stimulate Food Production (GCP/GLO/267/JPN) with contribution from the Ministry of Agriculture, Forestry and Fisheries (MAFF) of Japan. This document presents findings from its project activities.

The report seeks to build a better understanding of the relationship between savings and investment at the farm level, domestic and foreign corporate private investment in agriculture and agro-industries and the public sector investment in developing countries. It proposes policies and a programme of action for creating conditions under which domestic savings, farm-level investments and investment in agro-industries are united in a self-perpetuating, virtuous cycle that can be described as 'save, invest and grow'.

LAURENT THOMAS
Assistant Director-General
Technical Cooperation Department

Acknowledgements

This report was prepared by Saifullah Syed, Senior Economist, FAO Investment Centre (TCI), in collaboration with Masahiro Miyazako, Project Coordinator, TCI. We would like to express our gratitude to the Japanese Ministry of Agriculture, Forestry and Fisheries for funding this work.

The report is based largely on country case studies and a literature survey. It has drawn extensively from the FAO publication: *State of Food and Agriculture 2012 – Investing in Agriculture for a Better Future*. The case studies were conducted in Bangladesh, Bolivia, Brazil, Burkina Faso, Cambodia, China, Ethiopia, Egypt, India, Indonesia, Lao People's Democratic Republic, Mali, Malawi, Nepal, Republic of Korea, South Africa, Paraguay, Thailand, United Republic of Tanzania, Viet Nam and Zambia. We would like to thank the researchers and consultants who prepared the case studies and the background analytical reports.[1]

The report benefited from the Technical Workshop on Policies for Promoting Investment in Agriculture, held in Rome, 12–13 December 2011 and led by Peter Hazell, Imperial College, London, as a key resource person. It has also benefited from the Symposium on Poverty Reduction and Promotion of Agricultural Investment, held on 10 March 2010 in Tokyo, Japan, with the participation of Supachai Panitchpakdi, Secretary-General of the United Nations Conference on Trade and Development (UNCTAD) as well as the Workshop on Private Corporate Sector Investment in Agriculture in Southeast Asia, organized by

---

[1] Names of authors of the case studies are given in Annex 4.

Brighten Institute, Bogor, Indonesia in collaboration with FAO and held on 10–11 November 2012 in Bandung, Indonesia. We are grateful to all the participants to the workshops and symposium as well as the FAO Liaison Office in Japan and Brighten Institute in Indonesia.

Datasets on Agricultural Capital Stock and other statistical indicators were revised and updated through the efforts of Dominic Ballayan, Carola Fabi and Robert Mayo of the Statistics Division. We are thankful to Pascal Liu, Suffyan Koroma, Pedro Arias and Massimo Lafrate of the Trade and Markets Division for supervising and guiding the case studies on foreign investment in agriculture and for contributing in the sections dealing with foreign investment, particularly Section 7.4 of this report. We also extend our thanks to Masataka Fujita and Astrit Sulstarova of UNCTAD for their valuable support on the data and analysis of foreign investment in agriculture.

We are grateful to Calvin Miller, Emilio Hernandez and Nomathemba Mhlanga, the Rural Infrastructure and Agro-industries Division, for their contributions on investment in agro-industries. We are also grateful to Keith Wiebe, Jakob Skoet, Sarah Lowder and Brian Carisma, all from FAO's Agricultural Development Economics Division, and Daneswar Poonyth of FAO's Policy and Programme Development Support Division. Special thanks to Peter Hazell (Imperial College, London), Martin Evans, Derek Byerlee, Rita Butzer (University of Chicago), Carlos Seré (International Fund for Agricultural Development), Alberto Valdés (Universidad Católica de Chile), Mahendra Dev (Indira Gandhi Institute for Development Studies), Tewodaj Mogues, International Food Policy Research Institute, Bisaliah Siddinaik, Donato Antiporta and Nasredin Elamin (FAO Regional Office for the Near East and North Africa).

We would like to thank Richard China, Director, FAO Liaison Office with the European Union and Belgium and David Phiri, Principal Adviser with FAO's Economic and Social Development Department, for providing guidance on the implementation of the project in their previous capacity as Director, Policy and Programme Development Support Division and Chief, Policy Assistance Support Service, respectively.

Finally, we would like to thank Ms Rumiana Uzunova for handling the administrative and logistical matters, helping in the preparation of this publication and harmonizing all other project activities.

Abbreviations and acronyms

ACS	Agricultural Capital Stock
AOI	Agricultural Orientation Index
BDT	Bangladesh Taka
CFS	Committee on World Food Security
CIP	Country Investment Plan
CSO	Civil Society Organization
DRC	Domestic Resource Cost
FDI	Foreign Direct Investment
G8	Group of Eight
GATT	General Agreement on Tariffs and Trade
GDP	Gross Domestic Product
GDS	Gross Domestic Saving
GFCF	Gross Fixed Capital Formation
HDI	Human Development Index
IFAD	International Fund for Agricultural Development
IFC	International Finance Corporation
IFPRI	International Food Policy Research Institute
IMF	International Monetary Fund

IPA	Investment Promotion Agency
MAFF	Ministry of Agriculture, Forestry and Fisheries of Japan
MDG	Millennium Development Goal
NEPAD	New Partnership for Africa's Development
NGO	Non-governmental Organization
ODA	Official Development Assistance
OECD	Organisation for Economic Co-operation and Development
PER	Public Expenditure Reviews
PPP	Purchasing Power Parity
PRAI	Principles for Responsible Agricultural Investment that Respects Rights
R&D	Research and Development
RRA	Relative Rate of Assistance
SME	Small- and Medium-sized Enterprise
SOFA	The State of Food and Agriculture
SPEED	Statistics of Public Expenditure for Economic Development
TFP	Total Factor Productivity
UN	United Nations
UNCTAD	United Nations Conference on Trade and Development
UNDP	United Nations Development Programme
UNIDO	United Nations Industrial Development Organization
VGGT	Voluntary Guidelines on the Responsible Governance of Tenure of Land
WDI	World Development Indicator
WTO	World Trade Organization

Executive summary

As food prices increase and many face greater food insecurity, there is global concern about financing agricultural growth in the developing world. Accelerated agricultural growth is not only needed to meet growing global demand for food and cncrgy, but is also seen as the main pathway out of hunger and poverty for many impoverished people and countries.

Global estimates of the amount of investment needed to achieve acceptable levels of agricultural growth in the developing world vary enormously, but all these estimates far exceed the current trends in investment in agriculture by governments and donors. This report addresses the key questions that need to be addressed if this investment gap is to be filled:

- Who is going to make the required investments?
- Where are the needed resources going to come from?
- What are the policy options to ensure that resources flow to appropriate segments of the agrofood value chain?
- What are the policy options for guiding these investments to enhance pro-poor outcomes?

SOME NOTABLE FEATURES OF INVESTMENT IN AGRICULTURE

Investment in agriculture must be viewed in the wider economic context in which agricultural development occurs. In designing policies and programmes for promoting investment in agriculture it should be recognized that agricultural development depends on the simultaneous growth of agricultural production and the value chains to which it is linked. These value chains include a wide range of small- and large-scale activities that involve supplying farm inputs, processing, storing, distributing, wholesaling, retailing and exporting farm products. All these activities can be referred to collectively as 'agro-industry'. When considering agricultural investment, it should also be recognized that there are different types of investors operating in agriculture and its value chains. These investors have different objectives and roles. In addition, there is a variety of sources of financing for investment and all the sources are not equally accessible to all investors.

It must also be acknowledged that investment capital comes in many forms: financial capital, productive capital, fixed capital, working capital, as well as human capital, social capital and natural capital. Different forms of capital cannot simply be added together to determine the total amount of capital available or needed. They overlap and complement each other, and some forms of capital cannot be substituted for others. Moreover, different types of investors exercise varying degrees of control and ownership over these different types of capital, and at different stages along value chains. Investments made by different investors can on occasion complement each other, whereas in other situations different investments cannot simply be substituted for others. For example, farm capital formation, which is essential for increasing agricultural production, depends on farmers' own investment of labour and financial resources and cannot be substituted by other investors and sources of financing investment. Likewise, there are certain areas where only the public sector can or will invest. Conflicting goals can arise between different types of investors leading to tradeoffs that require public intervention to find the right balance between the economic, social and financial costs and benefits.

FARMERS ARE THE BIGGEST INVESTORS IN AGRICULTURE

For any investment to have positive impact on production and productivity, it must contribute to capital formation at the farm level. In this respect, the investments made by the farmers themselves are indispensable. Public investment in agriculture and private investment in agro-industries complement farm-level investment, but cannot substitute for the investments that need to be made by the farmers themselves.

The most comprehensive and readily available data for empirical measurement of investment in agriculture are the FAO estimate of on-farm capital stock. FAO has prepared estimates of on-farm capital stock for 206 countries from

1975 to 2005 based on inventories of agricultural assets contained in the FAOSTAT database. According to this dataset, farmers are the largest source of investment in agriculture for agricultural capital stock (ACS). On-farm investment in agricultural capital stock by farmers is nearly three times as large as other sources of investment combined, including public investment, foreign direct investment and official development assistance.

According to the FAO publication, *The State of Food and Agriculture 2012 – Investing in Agriculture for a Better Future* (FAO, 2012a), in the 47 countries that are on track to achieve the Millennium Development Goals (MDGs) hunger-reduction target, agricultural capital stock per agricultural worker (a proxy for private domestic agricultural investment) has grown by 0.7% per year since 1992. Whereas, this ratio has declined slightly in the 25 countries where progress has been insufficient and strongly in the 15 countries where rates of undernourishment have stagnated or regressed.

PUBLIC INVESTMENT

Available data indicate that public investment, although small relative to farmers' investment, is the second most significant contributor to farm-level capital formation, both directly through the provision of rural public goods and its effect on private investment. Hunger is more prevalent in countries where public agricultural expenditure per agricultural worker is lower, suggesting that both public and private investment in agriculture is important in the fight against poverty and hunger.

THE CORPORATE PRIVATE SECTOR AND FOREIGN DIRECT INVESTMENT

There are no comprehensive data on corporate private sector and foreign direct investment (FDI) in agriculture. However, the limited and country-specific data collected through case studies and sources such as UNCTAD, demonstrate that bulk of the corporate investment goes into agro-industries and the higher end of the value chain. Private sector investments along value chains are opening up new market opportunities for some farmers, but it is also becoming apparent that many small farmers are being left behind. There are signs of an increasing chasm opening up between small farms that are commercializable and non-commercializable subsistence farmers. This polarization could lead to a situation where policies and investments geared towards strengthening commercial agricultural production and value chains are not consistent with policies and investments geared to reducing poverty and food insecurity.

INVESTMENT REQUIRES THE RIGHT CAPACITIES AND THE RIGHT ENVIRONMENT

All agricultural investors, regardless of their size or the country, require the capacity to make investments and an environment that enables them to do so.

For farmers, their capacity to invest is determined by their main sources of investment finance: their own savings and their fixed capital, which is used as collateral for credit. Capital formation is certainly higher for farming households with positive savings. In the countries where the levels of poverty and hunger are high, the average farmer does not have any savings. In India and Bangladesh, more than 80% of the farming households demonstrate negative savings and take out loans just to cover their consumption. In recent years, remittances from migrating family members have contributed to increasing investment in agriculture. However, policies to provide credit to small and marginal farmers, who do not have adequate collateral, have not had the desired success.

Migration and remittances have recently become a main source of rural household income in many developing countries. They were found to be an important source of investment in agriculture for the development of family farming and particularly for making the shift from subsistence agriculture to market-oriented production. Migration is predominantly a family decision. It is the family that decides whom to send, mobilizes the cost of migration and, in return, receives remittances for the wider benefit of the family. However, it should be noted that large part of remittances are used for immediate consumption, health and education. Only a small proportion, around 10–12%, is invested in agriculture.

ESSENTIAL REQUIREMENTS FOR ENHANCING THE CAPACITY TO INVEST

There are several essential requirements for increasing savings and domestic investment in agriculture. They must work in tandem with each other and with sectoral and overall policies. Fulfilling only one of these requirements without considering the others is not likely to be effective in promoting investment.

Ensure ownership, transferability and transformability of capital through good governance and rule of law

People save to transfer and eventually transform their savings into capital. For this process to function efficiently, good governance and rule of law are required. To be effective, the legal system must be equally accessible and affordable to all.

Establish secure property rights, fixed capital and financial institutions

Fixed capital formation is a driving force for economic growth, development and the reduction of poverty and hunger. The crucial factors that allow for the formation of fixed capital are clearly defined property rights that are applied fairly and equitably to all under the rule of law and the presence of working financial institutions. Secure rights to land encourage investment, and financial institutions enable fixed capital to become a source of investment.

Allow and facilitate land consolidation to ensure a level of income that is adequate for savings

In most countries with acute food insecurity and poverty, most of smallholder farmers are not in a position to save. To promote farm-level investment, land consolidation needs to be facilitated to enable farmers to attain a level of income adequate for positive savings. Land consolidation, however, needs to be supported by an exit strategy for those who cannot make a living in agriculture, which generates non-farm income opportunities and provides appropriate social protection measures.

THE ENABLING ENVIRONMENT

The enabling environment for making investments depends on essential public goods, such as rural infrastructure, including roads and electricity, which farmers and the private sector cannot be expected to provide. This requires government action and supportive policies and institutions. Market forces that shape investment decisions, which are largely determined by the private sector, can also be influenced by government policies.

For enhancing the capacity of business to promote agro-industry investment, government can consider a hierarchy of enabling conditions. Essential enablers include land tenure and property rights, infrastructure, and domestic and foreign trade policy. Important enablers include norms, standards, regulations and services relating to production, research and development, and financial services for agro-industries. Useful enablers include the ease of doing business in a country, the business development services available to prospective investors, and the general intensity and effectiveness of business linkages between enterprises in value chains.

PUBLIC SECTOR INVESTMENT

Public investment is deemed to be exogenously determined in the sense that political economy compulsions determine the level and composition of public investment. Therefore, increasing attention is being paid to improving budget and the policy making process. The political economy considerations influencing agricultural policy choices includes, among other things, ideas and ideology. These play an important role in explaining agricultural policy choices.

In many countries, the public sector is making concrete efforts to guide and improve investment in agriculture by developing country investment plans (CIPs), based on predefined development strategies and national priorities. It is important that the development strategy leads to the adoption of policies and programmes that will contribute to increasing farm household savings and investment.

This report proposes a three-pronged strategy for pro-poor agricultural growth that involves:

i. Promoting the growth of commercial agriculture and its value chains, using public policy to enable the private sector (farmers and agro-industrialists) to take the lead.

ii. Shaping the engagement of the public sector in ways that enable as many small farms as possible to link to markets and successfully commercialize by investing their own savings.

iii. Putting in place support programmes targeted to those small farms that cannot succeed as viable businesses (e.g. facilitating exit strategies for those who cannot succeed in agriculture and enacting measures to promote rural non-farm employment).

SAVE, INVEST AND GROW

Farmers who cannot save cannot invest, and any economic activity that does not generate positive savings is not sustainable. Savings are essential not only for increasing the level of capital, but to cover the depreciation of the current level of capital stock. Farmers' capacity to invest depends on their capacity to save. This report proposes strategies to enhance farm household savings, leading to increased investment in agriculture. They are aimed at facilitating farmers' own efforts to increase their savings and investment in order to break away from the vicious cycle of poverty and enter the virtuous cycle that can be described as 'save, invest and grow'.

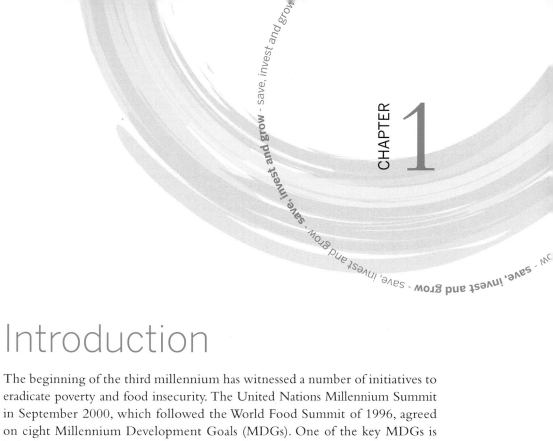

CHAPTER 1

Introduction

The beginning of the third millennium has witnessed a number of initiatives to eradicate poverty and food insecurity. The United Nations Millennium Summit in September 2000, which followed the World Food Summit of 1996, agreed on eight Millennium Development Goals (MDGs). One of the key MDGs is halving global poverty and hunger. The MDGs are part of a broader attempt to encourage the international community to join forces in making a difference in the developing world. Driven by these initiatives, development cooperation entered a phase of renewed growth and emphasis. The Organisation for Economic Cooperation and Development (OECD) and the Group of Eight (G8) countries made commitments to increase assistance to the developing world. During the United Nations Conference on Sustainable Development (Rio+20) in June 2012, the Zero-Hunger Challenge was launched, which calls for an end to world hunger.

The development concerns of developing countries also formed an integral part of the 2001 Doha Ministerial Declaration. Recognizing the fundamental principles of the World Trade Organization (WTO) and relevant provisions of the General Agreement on Tariffs and Trade (GATT) in 1994, the Doha Ministerial meeting agreed to pay special attention to the concerns of the developing countries. On the assumption that global food supply was sufficient to meet the global food demand, the WTO agricultural negotiation focused on how to improve market access of food-importing countries.

However, in 2008, soaring food prices changed the world food security situation. The crisis cast doubt on the belief that the global food supply was sufficient to meet demand. The introduction of export bans on food items in response to soaring prices also created severe hardship for poor food-importing countries. To address this critical situation, world leaders gathered in Rome in June 2008 for the High-Level Conference on World Food Security: The Challenges of Climate Change and Bioenergy. The world leaders recognized that reductions in food insecurity and poverty are positively related to overall economic development. They also recognized that due to its strong linkages with the other economic sectors, growth in agriculture is crucial. The Joint Statement on Global Food Security, which came out of the G8 meeting in L'Aquila, Italy in July 2009, acknowledged that consistent underinvestment in agriculture, combined with economic instability, were some of the main reasons for the persistence of food insecurity.

Every country that has made the transition to development, reduced poverty and increased food security has done so during periods of high agricultural growth. Empirical evidence shows that higher levels of economic development and non-farm activities are positively correlated with agricultural development, particularly with improved efficiency of the sector in terms of land and labour productivity and its aggregate value added. Conversely, the persistence of poverty and food insecurity is often associated with, and can largely be attributed to, lower growth of agriculture as well as low land, labour and total factor productivity.[2] The experience of developing countries strongly suggests that a sustained increase in agricultural production and productivity is required to make the transition from economic stagnation to self-sustaining growth in the agricultural sector and consequently in the overall economy.

The latest UN estimates suggest that by 2050 the world's population will have increased from 6.8 billion people to 9.1; a 34% increase over the next 41 years. FAO has estimated that agricultural production needs to grow by 70% over the same period to feed this population. This increased production is required because of a shift in demand towards higher value products of lower caloric content and a greater use of crop output as feed to meet the rising demand for meat. These estimates for additional output are likely to be low, as they do not take into account increases in agricultural production to meet the expanding demand for biofuels (FAO, 2009).

In the same study, FAO calculates that the investments needed in developing countries to support the required expansion in agricultural output far exceed the current trend. Another challenge is to increase capital stocks in areas that are lagging both in terms of hunger reduction and agricultural productivity.

---

[2] It should be noted that poverty and food insecurity are also associated with lower Total Factor Productivity (TFP), which includes technical progress and efficiency of resource use.

A study looking at the long-term record of investment in agriculture since the 1970s showed that, in general, the countries that performed best in terms of reducing hunger were also countries that manifested higher net investment rates per agricultural worker. Throughout the 1990s, in countries where less than 2.5% of the population was undernourished, the value added per worker was about 20 times higher than in countries where more than 35% of the population was undernourished.

In view of this, FAO, with financial support from MAFF, Japan initiated a project: Support to Study on Appropriate Policy Measures to Increase Investment in Agriculture and to Stimulate Food Production. The aim of the project is to identify a policy framework for promoting, facilitating and supporting the acceleration of investment by the public and private sector to achieve domestic capital formation for stimulating sustainable food production.

The process of formulating a policy framework for promoting investment requires a clear understanding of what conditions drive investment. Appropriate policies and measures must then be designed to promote and facilitate these conditions. This report identifies drivers of investment and then analyses policy options that cause those drivers to channel investment into agriculture.

Chapter 2 presents the concept and definition of investment in general, and Chapter 3 analyses investment in agriculture. Chapter 4 presents levels and trends of current investment in agriculture at the global and national level. Chapter 5 describes the different investors who invest for capital formation at the farm level and their relative contributions. It focuses on the private sector, the public sector, official development assistance (ODA) and FDI. Chapter 6 discusses the drivers of investment for farm-level capital formation and in agro-industry. Chapter 7 looks at ways to promote investment for on-farm capital formation, investment by the public sector, investment in agro-industry and FDI.

CHAPTER 2

The concept and definition of investment

2.1 WHAT IS INVESTMENT?

The term investment refers to forgoing consumption in the present to pursue a higher level of income in the future. Investments include: the purchase of stocks, shares, bonds and securities; the purchase or building of real property, such as residential or commercial land and/or real estate; and the purchase of machinery, equipment and transport for commercial purposes.

Farmers and governments invest in order to build capital, which allows the agricultural sector to become more productive in the future. Investment is generally defined as activities that result in the accumulation of capital, which yields a stream of returns over time. In economic growth theory, initiated 70 years ago by Harrod and Domar, investment is simply a change in capital stock or fixed inputs used in a production process (Harrod, 1939; Domar, 1946). From the 1940s to the present, the Harrod and Domar growth formula has been widely adopted and used for calculating target rates of investment in economic planning and development. In the words of Joan Robinson, 'By investment is meant an addition to capital, such as occurs when a new house is built or a new factory is built. Investment means making an addition to the stock of goods in existence' and it is the part 'of the production not merely replacing past sales, but is directed to increasing the rate of output in the future' (Robinson, 1956).

In official national accounts, investment is mainly referred to as gross fixed capital formation (GFCF), a macroeconomic concept. This concept does not make any adjustments to exclude the consumption of fixed capital (depreciation of fixed assets) from the investment figures. Regarding land, GFCF includes only the value of land improvement as a net addition to wealth. Investment is largely about changes in produced non-financial assets, the stock of which can be increased through economic activities. Annex 3 gives a full description of the non-financial assets in the System of National Accounts.

In the System of National Accounts 2008, apart from GFCF, investment includes: changes in stocks of inventories, including raw materials and final products; acquisition less disposal of valuables; depreciation; and acquisition less disposal of natural resources and third party property rights.[3]

2.2 DISTINCTION BETWEEN INVESTMENT AND EXPENDITURE

As elaborated in the State of Food and Agriculture (SOFA) 2012 (FAO, 2012a), some of the ways farmers invest in their farms include: acquiring farm equipment and machinery; purchasing animals or raising them to productive age; planting permanent crops; improving their land; and constructing farm buildings. Government investments include building and maintaining rural roads, and large-scale irrigation infrastructure. These assets generate returns in terms of increased productivity over a long period of time. Governments also invest in other, less tangible, assets such as the legal and market institutions that form part of the enabling environment for private investment. Determining whether expenditure, public or private, constitutes an investment can be both conceptually and empirically difficult. In some cases, the determination is not clear-cut.

In agriculture, a distinction is usually made between investments and spending on inputs. This distinction is based rather arbitrarily on the length of time required to generate a return. Planting trees is typically considered an investment because it takes more than a year to generate a return. However, applying fertilizer to a maize crop is not considered an investment because it generates a return during the immediate crop cycle. More important from a conceptual point of view, trees are a capital asset that yields a stream of returns over many years. Even in this seemingly

---

[3] FAO Statistics Division is developing a conceptual framework for an integrated Investment Dataset comprised of four main elements (Credit to Agriculture, Government Expenditure on Agriculture and Rural Development, Official Development Assistance to Agriculture and Foreign Direct Investment in Agriculture) as described in 'Statistics on Private and Public Investment in Agriculture' available at: www.fao.org/fileadmin/templates/ess/ess_test_folder/Workshops_Events/APCAS_24/Paper_after/APCAS-12-26_-_FAO_Investment_Dataset.pdf

simple case, the distinction between investments and spending on inputs may not be clear. If fertilizer use helps maintain and build soil fertility in the long run, it may also be considered an investment. Similarly, in public expenditures, a distinction is generally made between investment and current expenditures. Again, this distinction is not always clear-cut because current expenditures are required to maintain the value of capital assets, such as roads and other physical infrastructure.

Perspective also matters for what is perceived as investment. From a farmer's point of view, the purchase of land may represent an important investment in his productive capacity. From society's perspective, however, this purchase simply involves a change in ownership of an asset rather than a net increase in capital stock, which occurs for instance when land improvements are undertaken.

Investment therefore is a flow and involves the formation of capital. It does not represent the stock of capital in an economy, but rather the changes in that stock of capital that are intended to increase future production, output or income. If it is accepted that the general definition of investment is the increase of capital goods in a given period of time, then the next question to be asked is: what is capital?

2.3 WHAT IS CAPITAL?

The term capital means purchasing power or a fund of generic wealth, owned by individuals or firm and destined to earn its return. In everyday speech the linkage between possession of capital and attainment of a return is emphasized, but the devil is in the details. In general, the definition of capital is a group of 'products that serve towards production' or as groups of 'produced means of production'. This excludes products that serve for immediate satisfaction of needs, as well as land, since it is not a produced item.

However, it is difficult to conceptualize capital for productive investment because it is a diverse set of physical items, such as plants, machinery, buildings, tools and vehicles that are used in the production process. Capital includes human-made goods (or means of production) used in the production of other goods and services. It includes physical items of different kinds and ages, with different technological content (and different levels of obsolescence), which are not conceivable as a homogenous group. It is, however, possible to make the distinctions between fixed (or physical) capital and working capital.

In addition to the above classification of capital, the French sociologist Pierre Bourdieu proposed another distinction of various types of capital. According to Bourdieu, capital acts as a social relation within a system of exchange to extract profits. It can be divided into different categories: economic capital (command over economic resources, such as cash and assets); social capital (the aggregate of the actual or potential resources that are linked to possession of a durable network of more or less institutionalized relationships of mutual acquaintance and recognition and based on group membership, relationships, networks of influence and support); cultural capital (forms of other non-financial social assets, such as knowledge, skills, education, and advantages that a person receives from their parents and educational system to promote social mobility beyond economic means and obtain a higher status in society); and symbolic capital resources available to an individual on the basis of honour, prestige or recognition (Bourdieu, 1986).

These different forms of capital cannot simply be added together to determine the total amount of capital available or needed. They overlap and complement each other, and some forms of capital cannot be substituted for others. Also, all forms of capital are not equally important for agriculture and/or for the different stages of the food value chain. Before discussing the promotion of investment in agriculture, it is important to be clear about what kinds of capital are relevant for agriculture and take into account the wider economic context in which agricultural development occurs.

CHAPTER 3

Investment in agriculture for increased production and productivity

Agricultural development depends on the simultaneous growth of farm-level production and productivity and the value chains to which it is linked. These value chains include a wide range of small- and large-scale activities that involve supplying farm inputs, processing, storing, distributing, wholesaling, retailing and exporting farm products. These activities can be referred to collectively as 'agro-industry'. There is a need to look at both farm-level investment, as well as investment in agro-industries.

3.1 INVESTMENT FOR ON-FARM AGRICULTURAL CAPITAL STOCK

For any investment to have positive impact on production and productivity, it must contribute to capital formation at the farm level. Persistent poverty and food insecurity is partially explained by insufficient food production, due primarily to the low productivity of agriculture. Low productivity of agriculture signifies low per unit output of factors of production. The primary factors of production in agriculture are land and labour. Low labour productivity, or low land productivity, or the combined effects of both are accountable for low productivity of agriculture.

Agricultural products are outcomes of tamed natural processes that take place on land. This is accomplished with human labour. Land has to be worked to generate agricultural outputs. Land and labour are indispensable primary factors. Without them, agriculture does not exist. Land, as non-produced asset, is fixed in supply. Labour is inherently variable. The labour force or the amount of time worked can change depending on the population or on workers' preferences. Agriculture on the aggregate level is an industry characterized by a combination of fixed land with variable capital. This combination is typical of low-productivity agriculture in which land size has natural limits and the agricultural labour force is expanding. For the economy as a whole, there is little scope to increase the expanse of agricultural land, particularly when concerns for the environment are mainstreamed into economic activities. Increasing numbers of people taking refuge in agriculture for their livelihoods is a fact of life in developing economies. As such, the model of agriculture relevant to this analysis is characterized by fixed land and variable labour, which are determined by environmental, socio-economic, political and demographic factors that lie outside agriculture.

According to economic theory, a fixed tract of land combined with increased labour produces increased output at a decreasing rate, as it is bound to face inescapable diminishing returns to labour. Productivity increases from the land decline as the land is worked with more labour. Moreover, land loses its fertility as soil nutrients get extracted through repeated cropping. A model of fixed land with increasing labour and declining fertility loss through time approximates the reality of underdeveloped agriculture. This reality will not change unless a compensating mechanism is put in place in the form of land improvement and fertilizing. Indeed, underdeveloped agriculture is characterized by the low level of compensating mechanisms to offset fertility loss. The set of compensating mechanisms is part of a broader concept called land-augmenting technology (Todaro and Smith, 2003), which is the application of a certain form of capital that enhances the productivity of land. That form of capital consists of newer methods and newer technologies for doing things.

Raw agricultural labour is the human expenditure of energy to do useful work. There are natural limits on the amount of energy an individual worker can exert. The natural limit of raw human labour can be extended with tools and mechanical implements. However, agricultural labour is not simply an application of physical force. In working the land and in dealing with natural forces, knowledge and skill guide the physical labour that is augmented by the use of tools and implements. Tools, implements, skill and knowledge are all factors of production that constitute a form of capital that enhances the productivity of labour. It can be referred to as labour-augmenting capital. Literature makes a clear distinction between capital and technology. However, empirically it is difficult to separate technology and capital. In addition, the ultimate rationale for the use of technology in combination with capital is to augment labour. For these reasons, it is unnecessary to expend effort in distinguishing between capital and technology. The use of capital fused with technology may be conceptualized as labour-augmenting capital in the same way that there is land-augmenting capital.

In developing country agriculture, modern skills, knowledge, tools and implements are rudimentary, which leads to low productivity. An expanding agricultural work force with rudimentary skills, knowledge, tools and implements creates a situation where output per worker (agricultural labour productivity) declines. This is the result of a combination of two factors: a growing number of farm workers and inadequate working capital.

The possible combinations of raw labour, land, capital-augmented labour and capital-augmented land provide a range of possibilities for productivity. The lowest level of productivity (traditional agriculture) arises from the combination of raw labour and land. The highest productivity (modern, developed agriculture) results from the combination of capital-augmented labour and capital-augmented land (see Fig. 1).

FIGURE 1.
Productivity and combinations of land and labour with and without capital

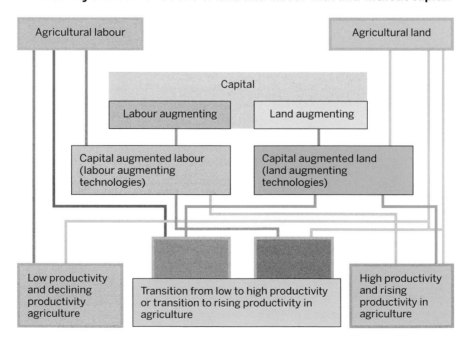

In underdeveloped agriculture, where declining soil fertility is not compensated by investments in land improvement, fertilizers and chemicals, and where the expanding labour force is not equipped with modern skills, knowledge, tools and implements, the productivity of both labour and land is bound to decline. Productivity of labour declines as more households with low labour-augmenting capital become dependent on agriculture. Combining increased amounts of labour inputs with fixed tracts of land would inevitably lead to diminishing returns to labour. Diminishing returns alone do not lead to declining productivity. Losses in soil moisture and fertility caused by environmental degradation and the depletion of nutrients from repeated cropping also cause productivity to decline. Without the application of land-augmenting capital, farm land depreciates and the productivity of the land declines, which leads to more exploitation of land and the environment. Traditional agriculture uses raw human labour combined with traditional wisdom and rudimentary forms of capital. Draught animals and unsophisticated farm implements and tools are predominant. Traditional agriculture in areas with an expanding population is an example of a situation in which there is an increasing variable factor (labour) combined with weak labour-augmentation capital and declining (rather than fixed) production factors (soil, land and other natural assets). Productivity loss is accentuated by both increasing raw labour and declining soil fertility and environmental quality.

Productivity losses due to increases in raw labour, declines in soil fertility and environmental quality can be reversed by introducing modern skills to farm labour (formation of human capital in agriculture), applying labour-augmenting technologies compatible with the skills of the labour force (investment in physical capital in agriculture), integrating land-augmenting technologies and improving the quality of natural assets (physical capital formation in and for agriculture). Such inputs offset the effects of diminishing returns and the depreciation of land and natural assets. Accelerated capital formation in agriculture reverses the process of declining labour and land productivity.

The significance of capital formation in agriculture for growth is evident from the fact that investment in machinery and equipment enables the farm worker to work on larger tracts of land and make use of other forms of capital, such as livestock. The investment in machinery and equipment augments labour, renders it more productive and offsets the effects of diminishing returns. If land and other inputs are fixed, investment in land-augmenting inputs counters the effects of land depreciation and diminishing returns by increasing yields. The importance of capital formation for agriculture is demonstrated when investments in service and industrial sectors (education, transport, manufacturing facilities for fertilizers and pesticides) succeed in enhancing agricultural productivity. Investments in other economic sectors outside agriculture and its related activities also represent, to a varying extent, investments in agriculture.

3.2 INVESTMENT BY THE PUBLIC SECTOR

In addition to capital formation at the farm level, agriculture requires public expenditures for public goods, including rural public goods. Investment in rural public goods, such as education, infrastructure, health care and social services, can generate important benefits for the agricultural sector and boost its contribution to economic growth and poverty alleviation. Rural public goods are complementary to private farm-level investment; investing in one often enhances investments in the other. Evidence also shows that agricultural productivity and poverty reduction are compatible goals, with investments in rural public goods usually having high payoffs for both (FAO, 2012a).

Recent studies have highlighted that public expenditures on agricultural research, education and roads contribute strongly, but to varying degrees, to agricultural growth across regions. The impact of research expenditures on productivity is stronger than non-research expenditures (Fan and Saurkar, 2006). Investment in research, often associated with extension, is consistently found to be the most important driver of productivity growth in agriculture (Fischer *et al.*, 2009).

Studies have compared public spending on agriculture with other forms of expenditure as to the impact they have on agricultural performance and poverty. After agricultural research and development (R&D), the ranking of returns to investment in other areas differs by country. This suggests that public investment priorities depend on local conditions. Nevertheless, rural infrastructure and road development are often ranked among the top drivers of overall economic growth in rural areas (Fan *et al.*, 2000, 2004; Mogues, 2011). In Ethiopia, access to all-weather roads reduced poverty by 6.9% and increased consumption growth by 16.3% (Dercon *et al.*, 2009). Mogues (2011) found that in Ethiopia, out of all forms of investment, the returns on public investment in road infrastructure were by far the highest. In Uganda, the marginal returns for public spending on feeder roads on agricultural output and poverty reduction was three to four times larger than the returns to public spending on larger roads (Fan and Zhang, 2008).

Public goods in rural areas also tend to be complementary. For example, in Bangladesh, villages with better infrastructure witnessed higher farm-level investments than villages with poorer infrastructure. These villages irrigated, used improved seeds and fertilizer, paid lower fertilizer prices, earned higher wages and had significantly higher production increases (Ahmed and Hossain, 1990). In Viet Nam, rural roads fostered the development of local markets and raised primary school completion rates, which created greater incentives for agricultural investment and increased investment in human capital (Mu and van de Walle, 2007).

3.3 INVESTMENT IN AGRO-INDUSTRY FOR AGRICULTURAL DEVELOPMENT AND GROWTH

As countries develop, agricultural production has to be increasingly marketed and processed to supply rapidly urbanizing populations and industries. With rising incomes and urbanization, people also diversify their diets into a wider range of higher value foods, including fresh perishable foods and processed and pre-cooked foods. Farmers also gradually adopt more capital-intensive farming methods to supply these growing markets. This process tends to be associated with greater access to modern inputs such as fertilizers, pesticides, improved seeds and machines. Agro-industry plays an important role in all these areas. As such, growth in agricultural investments and production depends on sufficient complementary investments in agro-industry. This fact is often ignored in discussions about the investment needs of agriculture.

In most developing countries, agro-industries are usually at the vanguard during the early stages of industrialization. Agro-industrial enterprises come in various forms. At one end of the spectrum, are small family-owned and operated rural businesses that are unregistered and officially unrecognized. At the other end, are the global transnational corporations listed on international stock exchanges. The growth and development of small- and medium-sized enterprises (SMEs) is in particular need of stimulation, rather than that of large enterprises. Many smaller agro-industrial enterprises are found in the informal sector. Their larger counterparts are invariably part of the corporate sector.

Agriculture and agro-industry are linked by their demand for outputs and by their capital stocks and new investments. These linkages are described below with the aid of some simple equations.

Agricultural output (Q_{ag}) depends on: the amounts of direct inputs used (XAg), which includes labour, land, irrigation, seeds, feeds and agrochemicals, such as fertilizers and pesticides; the choice of technology and management practice (Tech); the stocks of farm-level capital held by farmers (KAg); the capital stock held by agro-industry (KAgIn); the stock of public sector rural capital (KPubR);

and the farm size distribution (FSD). Other variables may come into play, but they do not need to be listed here.

$$Q_{ag} = f(XAg, Tech, KAg, KAgIn, KPubR, FSD)$$

On-farm capital includes: productive trees, breeding animals, tools and machinery, buildings and storage facilities, land improvements (including irrigation infrastructure), human capital and rural social capital. The stock of private capital in agro-industry includes: trading establishments, warehouses, cold storage, processing facilities, transport equipment and human capital. Public sector rural capital includes: rural roads, public irrigation structures, rural electrification, research stations, rural schools and health centres. In each case, the composition of the capital stock is important, not only the stock's total value. In fact, several types of capital cannot be substituted for each other, for example human and land capital in agriculture.

Agro-industry output (Q_{ai}) depends on the amount of agricultural output (Q_{ag}); other direct inputs, including labour, fuel and purchased intermediaries (XAgIn); the stock of private capital in agro-industry (KAgIn); and the stock of public sector rural capital (KPubR). Again, there may be other variables, but they do not need to be enumerated here.

$$Q_{ai} = f(Q_{ag}, XAgIn, KAgIn, KPubR)$$

Within these relationships, agro-industry affects agricultural output through its supply of direct inputs, and its investment in agricultural R&D, which makes new technologies available to farmers. It also provides the main marketing channels for farmers and has an important bearing on the prices they receive and their opportunities to add value to their production. In turn, the level of agricultural output, the technologies and management practices used by farmers determine the demand for farm inputs and marketing, as well as storage and processing services.

The two sectors are also linked through their capital stocks. Investors in farm-level capital, for example, are not likely to invest much in building up their on-farm capital if they do not have adequate access to value chains for obtaining key inputs and marketing their products. Similarly, agro-industrialists are unlikely to build up their capital stock if they do not have access to sufficiently capitalized and market-oriented farmers that can create viable markets for their own outputs.

Therefore, investment by agro-industrialists is essential for:

- **Creating** agricultural value chains by linking farmers to suppliers on the one side and consumers on the other, usually through a series of intermediaries. In the absence of such linkages, farming will remain a low-productivity subsistence activity, largely isolated from the market economy.

- **Growing** value chains by connecting more farmers to (more) suppliers and (more) consumers or by increasing the volume or diversity of products handled.

- **Improving** value chains by making them more cost-efficient, principally by reducing loss and wastage and enhancing quality (including safety and traceability in the case of foods) as products move along the supply chain, and by exploiting economies of scale by moving higher and more stable volumes of produce. Agricultural value chains can be 'improved' by increasing the participation of smaller and poorer enterprises, including farms; by making participation by both farmers and agro-industrial enterprises more stable over time; and by ensuring that all participants behave in a more socially and environmentally responsible manner (Barrett *et al.*, 2012).

CHAPTER

4

Empirical measurement of investment in agriculture: the evidence from available data and information[4]

4.1 LEVEL AND TREND OF INVESTMENT FOR ON-FARM AGRICULTURAL CAPITAL STOCK (ACS)

The most comprehensive and readily available data for empirical measurement of investment in agriculture are the FAO estimates of on-farm capital stock. FAO has prepared estimates based on inventories of agricultural assets in the FAOSTAT database of on-farm capital stock for 206 countries from 1975 to 2005. There is no other database that comes close to it in terms of country and temporal coverage. The inventories-based estimates of ACS were first developed as an indicator for the *World Agriculture: Towards 2010* report (Alexandratos and FAO, 1995). The estimates were subsequently recalculated and improved in FAO (1999), FAO (2002), Barre (2006), and recently in FAO (2011). The estimates are based on data on the quantities of agricultural assets, which have been valued and

---

4   This section is devoted to analysis of on-farm agricultural capital stock. Data on investment in agro-industries for cross country analysis, as well as for estimating the level and trend of investment, are not available.

aggregated based on fixed US dollar prices of 2005. The estimates represent an index of quantities of ACS and are referred to as the inventories-based estimate of ACS. The indicator is composed of the following four main components: land improvements, livestock, machinery and farm structures. With regard to land, only the change in value due to improvement, or change in use, such as a transformation from grassland to arable land is considered an improvement. Data on machinery and livestock are obtained from FAOSTAT. In the absence of information on physical stocks, the value of hand tools is assigned a fixed value per agricultural worker. Also, to include farm structures, it is assumed that a fixed proportion of livestock (animal specific) is associated with a number and value of infrastructure. Depreciation (for land improvements, machinery and structure), which varies by component, is accounted for in the estimates of net ACS.[5]

According to this dataset, globally the volume of on-farm ACS has shown a very modest annual growth rate of 0.6% since 1980. It grew by 1% annually during the 1980s, followed by significantly lower growth rate of 0.2% during the 1990s. The growth rate recovered to 0.6% for the period 2000–2007.

TABLE 1.
Average annual growth in ACS, 1980–2007 (inventories-based estimates)

	AVERAGE ANNUAL RATE OF GROWTH %			
	1980/90	1990/00	2000/07	1980/07
World	0.98	0.18	0.60	0.58
High-income countries	0.50	-0.23	-0.09	0.07
Low- and middle-income countries	1.28	0.42	0.98	0.88
Region				
East Asia & Pacific incl. China	2.15	1.79	1.40	1.82
Of which: China	1.90	1.74	1.15	1.65
East Asia & Pacific, excl. China	2.63	1.90	1.85	2.16
Europe and Central Asia	0.48	-2.65	-0.58	-0.96
Latin America and the Caribbean	0.90	0.50	0.89	0.75
Middle East and North Africa	2.86	1.32	1.29	1.88
South Asia incl. India	1.43	1.46	1.34	1.42
Of which: India	1.44	1.54	1.10	1.39
South Asia, excl. India	1.40	1.34	1.73	1.47
Sub-Saharan Africa	1.30	1.51	1.89	1.53

SOURCE: Computed based on FAO, 2012a.

---

[5] More details about this dataset, as well as limitation of the other sources of data and information are given in: *Level, Trend and Sources of Financing for Investment in Agriculture: A Review and Analysis of Available Sources of Data* (FAO, 2012b).

However, the rate of growth has been uneven across regions and countries. A breakdown of the data between low- and middle-income countries and high-income countries (Table 1) shows that, in the low- and middle-income countries the average annual growth rate of ACS was 1.3% in the 1980s and declined to 0.4% in the 1990s. It increased to 1.0% during the period 2000–07. In high-income countries it grew at a lower rate, 0.5% in the 1980s, and declined in the 1990s and between 2000 and 2007.

At the regional level, the rates of ACS growth have been consistently positive across regions, with the exception of Europe and Central Asia. Sub-Saharan African countries had increasing average annual growth. South Asia, including India, had positive growth at fairly steady rate. East Asia and Pacific region as well as the Middle East and North Africa had decreasing average annual growth rates. In Latin America and the Caribbean, the growth rate decreased in 1990s and recovered between 2000 and 2007.

4.2 AGRICULTURAL CAPITAL STOCK PER AGRICULTURAL WORKER

ACS per agricultural worker is an important indicator because the amount of capital used by farmers or agricultural workers (the capital–labour ratio) largely determines labour and land productivity.

According to FAO data, from 1980 to 2007 capital stocks per worker at the global level decreased at an average annual rate of 0.5%. This decline is the result of the agricultural labour force expanding more rapidly than agricultural capital stock. In the low- and middle-income countries, the capital–labour ratio had a negative growth rate of -0.3%. In the high-income countries, the ratio of capital to labour increased, not as a result of a rapid increase in capital, but because of a rapid decline in the agricultural labour force (Table 2).

In terms of variation across geographical regions, per capita levels were lowest in East Asia and the Pacific, South Asia and sub-Saharan Africa. Per capita levels were highest in Latin America and the Caribbean, the Middle East and North Africa. The recorded decline in ACS per worker in the low- and middle-income country group may be due to the declining levels in sub-Saharan Africa and South Asia excluding India. Sub-Saharan Africa and South Asia excluding India, both saw an increase in total ACS over time, but not at a rate sufficient to keep pace with growth in the agricultural labour force. While capital–labour ratios increased in the remaining low- and middle-income regions, the highest average rates of growth occurred in the Middle East and North Africa, followed by Latin America and the Caribbean, and East Asia and the Pacific. In the Middle East and North Africa, and East Asia and the Pacific, the high growth in capital–labour ratios was a result of high rates of growth in overall ACS. In Latin America and the Caribbean, it appears that it was largely the result of slower growth in the agricultural labour force from 1980 to 2007. In Latin America and the Caribbean, capital–labour ratios have grown steadily. In the region, growth rates were higher during the 1980s and 2000s, but there was a significant slow-down during the 1990s.

TABLE 2.
Level and growth of ACS, (inventories-based estimates)

	Average ACS per agricultural worker 2005/07 (US$ 2005 constant)	AVERAGE ANNUAL RATE OF GROWTH %		
		ACS	Agricultural worker	ACS per agricultural worker
World	4,000	0.6	1.1	-0.5
High-income countries	89,800	0.2	-2.9	3.0
Low- and middle-income countries	2,600	0.9	1.2	0.3
Region				
East Asia and the Pacific	1,300	1.8	1.1	0.7
East Asia and the Pacific excl. China	2,000	2.1	1.4	0.7
Europe and Central Asia	19,000	-1.0	-1.7	0.7
Latin America & Caribbean	16,500	0.7	0.0	0.7
Middle East and North Africa	10,000	1.8	0.9	0.9
South Asia	1,700	1.4	1.4	0.0
South Asia excluding India	3,000	1.4	1.6	-0.1
Sub-Saharan Africa	2,200	1.5	2.1	-0.6

SOURCE: Computed based on FAO, 2012a.

Three studies (FAO, 2001; Schmidhuber *et al.*, 2009; von Cramon-Taubadel *et al.*, 2009) looked at ACS and the productivity of labour engaged in agriculture and its relation to the prevalence of hunger in developing countries. The findings of these studies reinforce the argument that ACS tends to have a positive impact on productivity of labour in agriculture and thereby a negative impact on the prevalence of hunger in developing countries. These three studies found that:

- There is a tendency for low levels of ACS per worker, low labour productivity and the high prevalence of undernourishment in the population to 'co-exist'.

- The regions with low capital intensity per agriculture worker also had low productivity per worker.

- The value added per worker in the group of countries with the lowest percentage of undernourished people was 2.0 times higher than in the group of countries with the highest levels of undernourishment.

- Total-factor productivity (TFP) growth rate in agriculture between 1975 and 2007 was the highest (2.1%) in China and lowest (0.9%) in sub-Saharan Africa. It was 1% in Latin America. (Details on decomposition of TFP into efficiency and technical change components are found in von Cramon-Taubadel *et al.*, 2009.)

To understand the relationship between per capita ACS, poverty and agricultural productivity, country case studies were carried out for a group of selected countries from Asia, Africa and Latin America. The selected countries are: Bangladesh, Burkina Faso, Ethiopia, Indonesia, Malawi, Malaysia, Mali, the Plurinational State of Bolivia, Republic of Korea, South Africa and Zambia. Table 3 gives some of the socio-economic indicators for the selected countries.

TABLE 3.
Socio-economic indicators

Country	Agriculture, value added (% of GDP)[1] 2000–07	GDP per capita (constant 2000 US$)[2] 2000–07	Percentage of under-nourishment in total pop.[3] 2000–08	HDI[4] ranking 2007
Ethiopia	46.0	142	44.5	171
Malawi	34.9	151	28.5	160
Burkina Faso	33.3	236	10	177
Mali	37.4	242	15	178
Zambia	22.5	346	43.5	164
Bangladesh	21.8	415	28	146
Indonesia	14.5	872	14	111
Bolivia (Plurinational State of)	14.6	1,049	24.5	113
South Africa	3.3	3,280	n.a.	129
Malaysia	8.9	4,342	<5*	66
Republic of Korea	3.7	13,129	<5*	26

SOURCES: [1] WDI; [2] WDI; [3] FAOSTAT; [4] UNDP.
Note: The Human Development Index (HDI) is a summary measure of human development. It measures the average achievements in a country in three basic dimensions of human development: a long and healthy life (health), access to knowledge (education) and a decent standard of living (income).
*MDG Indicator, 2000–06

Table 4 below shows the agricultural capital stock per worker in the countries where the case studies were conducted. Malawi has the lowest level of capital stock to labour ratio, followed by Burkina Faso and Ethiopia. South Africa has the highest level of ACS to labour ratio. Its level in 2007 was 2.7 times more than Malaysia and 3.1 times more than Republic of Korea. The ACS to agricultural labour ratio is very small in Malawi, Burkina Faso and Ethiopia. These countries have low per capita income and high levels of poverty, compared to South Africa, Republic of Korea and Malaysia.

TABLE 4.
Agricultural capital stock per agricultural worker (in constant 2005 US$)

	1980	1990	2000	2007
Ethiopia	-	-	1,363	1,649
Malawi	633	554	630	680
Burkina Faso	1,160	1,396	1,525	1,610
Mali	4,050	3,409	4,163	4,395
Zambia	2,498	2,196	1,987	1,970
Bangladesh	1,750	1,653	1,787	2,022
Indonesia	1,575	1,737	1,770	1,944
Bolivia (Plurinational State of)	6,414	5,148	4,876	4,931
South Africa	26,692	26,524	29,251	33,178
Malaysia	6,623	9,620	11,174	12,453
Republic of Korea	1,132	2,696	6,454	10,739

SOURCE: FAO, 2012a.

South Africa had the highest average level of ACS–labour ratio, but the highest annual average growth rate was in the Republic of Korea. In the Republic of Korea, the ACS to agricultural labour ratio grew by 8.7% between 1980 and 2007. Malawi, Mali and Bangladesh had a negative growth rate in between 1980 and 1990, but it was positive in the 1990s and between 2000 and 2007 (Table 5).

TABLE 5.
Average growth rate of agricultural capital stock per agricultural worker

	1980/90	1990/00	2000/07	1980/07
Ethiopia	-	-	2.8	-
Malawi	-1.3	1.3	1.1	0.3
Burkina Faso	1.9	0.9	0.8	1.2
Mali	-1.7	2.0	0.8	0.3
Zambia	-1.3	-1.0	-0.1	-0.9
Bangladesh	-0.6	0.8	1.8	0.5
Indonesia	1.0	0.2	1.3	0.8
Bolivia (Plurinational State of)	-2.2	-0.5	0.2	-1.0
South Africa	-0.1	1.0	1.8	0.8
Malaysia	3.8	1.5	1.6	2.4
Republic of Korea	9.1	9.1	7.5	8.7

SOURCE: Computed based on FAO, 2012a.

The results indicate that countries with higher growth of ACS per agricultural worker had higher per capita GDP and a lower prevalence of undernourishment in the total population.

CHAPTER 5

Who invests for farm-level capital formation?

5.1 INVESTMENT BY THE PUBLIC AND THE PRIVATE SECTOR FOR ON-FARM AGRICULTURAL CAPITAL STOCK

Investment is generally measured as incremental change in capital stock (ΔK) from one period to other. The value of the yearly depreciation of agricultural capital that needs to be replaced plus the yearly change is a proxy for estimating investment. The following equation $I_t = K_t - K_{t-1} (1 - \delta)$ is used to compute investment using inventory-based capital stock. In this equation (K_t) is the current agricultural capital stock; (K_{t-1}) is the previous year's capital stock; (δ) is the capital depreciation rate of 5%[6] (i.e. 5% of the ACS in the previous period that needs to be replaced); (I_t) is the current year's capital investment. It is equal to ($I_{t\,public}$) plus ($I_{t\,private}$). Public investment can be calculated as:

$$I_{t\,public} = I_t - I_{t\,private}; \text{ and private investment as: } I_{t\,private} = I_t - I_{t\,public}.$$

Many countries report internationally comparable data on public expenditure in agriculture. However, they do not distinguish between current expenditure and investments and information on public capital stock and investment flows. Systematic information on government expenditures does exist, but it is generally not possible to determine to what extent the reported expenditures can be considered investments that contribute to the capital formation.

---

[6] The rate of depreciation of different assets of 5% is problematic because, in terms of assets depreciation, not much information is available.

The most comprehensive databases that permit an assessment of government expenditures in agriculture and other sectors are: the Statistics of Public Expenditure for Economic Development (SPEED) database compiled by the International Food Policy Research Institute (IFPRI); and the FAOSTAT database on Government Expenditure in Agriculture.[7]

The FAOSTAT database on Government Expenditure in Agriculture contains data on all non-repayable payments, whether capital or current, and whether they are required by government or not. Expenditures are displayed by function or purpose at different levels of government administration: central government, state, region or province and local governments. The data refer to the share of expenditure on agriculture, forestry, fishery and hunting in the total government expenditure.

SPEED is the most comprehensive database on public expenditure. It includes, among other things, public expenditures on salaries and purchases of goods and services that do not contribute to capital formation. For this reason, the entire volume of public expenditure cannot be counted as investment.

How much of public expenditure can be considered as capital investment is a more empirical issue. According to the public expenditure reviews (PER) of

TABLE 6.
Share of government expenditures in agriculture that represents investment in agricultural capital

Country	Capital share of agricultural expenditures (%)	Year
Ghana	17	2005
Kenya	30	2004/5
Mozambique	84	2007
Nigeria	44	2001–2005
United Republic of Tanzania	9	2011
Uganda	24	2005/6–2008/9
Zambia	24	2000
Lao People's Democratic Republic	84	2004/05
Nepal	46	1999–2003
Philippines	26	2005
Viet Nam	77	2002
Honduras	66	2006
Simple average	44	

SOURCE: Lowder *et al.*, 2012.

---

[7] This database is not publicly available.

selected countries (Table 6), there is significant difference in the share of capital expenditures in total expenditures, ranging from as little as 9% in the United Republic of Tanzania to 84% in the Lao People's Democratic Republic and Mozambique. The average share of capital expenditures in total expenditures in the countries for which PER is available is about 44%. Using this average as the mid-point, it can be assumed that 40–50% of public expenditure could be considered as investment for capital formation.

Table 7 shows the proportion of total investment made by the private and public sector. It assumes that 50% of public expenditure is capital investment. Estimates were done for 76 low- and middle-income countries, whose public expenditure data are available in SPEED. The figures show that the bulk of the investment for capital formation is made by the private sector.

TABLE 7.
Sources of investment for capital formation at farm level, assuming 50% of public expenditure as capital investment (%)

Region/Country groupings	Sources of capital investment	Average 2005–07 or most recent year
East Asia and Pacific (12)	Public	40
	Private	60
Europe and Central Asia (12)	Public	19
	Private	81
Latin America and Caribbean (13)	Public	11
	Private	89
Middle East & North Africa (9)	Public	28
	Private	72
South Asia (7)	Public	13
	Private	87
Sub-Saharan Africa (23)	Public	10
	Private	90

SOURCE: Computed using data of Lowder *et al.*, 2012.

5.2 SOURCES OF INVESTMENT FOR CAPITAL FORMATION IN SELECTED COUNTRIES

For the selected countries, levels of public and private investment as percentage of total investment were calculated using the same methodology. Again, it was assumed that 50% of public expenditure is capital investment (Table 8). The results show that the bulk of the investment for capital formation is made by the private sector, that is, by the farmers themselves. The share of private investment ranges from 40% in the Republic of Korea to more than 99% in the Plurinational State of Bolivia.

TABLE 8.
Share of private investment in on-farm investment, assuming 50% of public expenditure as capital investment (%)

Country	1981/90	1991/00	2001/07
Bangladesh	92.2	89.2	90.1
Bolivia (Plurinational State of)	99.5	99.4	99.0
Ethiopia	99.0	98.6	98.8
Indonesia	96.4	95.4	98.1
Republic of Korea	50.7	57.5	40.1
Malawi	98.3	98.3	99.1
Zambia	88.3	98.5	94.3

SOURCE: Computed based on FAO, 2012b.

From the above table it can be seen that public sector is a small contributor to on-farm capital formation. However, it should be noted that public expenditure does not necessarily contribute directly to farm-level capital formation. Public expenditure contributes to increased production by providing the necessary public goods. In this context, farm-level investment by farmers and public investment are complementary and neither one can serve as a substitute for the other.

It is widely argued that one of the reasons for insufficient agricultural growth in many countries in recent years has been the relatively low levels of investment in essential public goods for agriculture. While real public spending on agriculture and spending per worker has grown in most regions, the share of agriculture in total public expenditure has not always kept up with its share in income generation. This is indicated by the agricultural orientation index (AOI) of public expenditure (Table 9). The AOI is defined as the share of agriculture in public expenditure divided by the share of agriculture in GDP. It is an indicator of the degree to which the share of agriculture in public expenditure is commensurate with the weight of the sector in GDP. Time trends in the AOI vary among regions. There has been a significant increase over time in East Asia and the Pacific, Eastern Europe and Central Asia, and South Asia, but declines in Latin America and the Caribbean, the Middle East and North Africa and sub-Saharan Africa. South Asia and sub-Saharan Africa, the two regions with the highest incidence of hunger and undernourishment, have low levels of agricultural expenditures per worker and of the AOI. In spite of recent commitments to increase agricultural expenditures, this situation is particularly pronounced in sub-Saharan Africa.

TABLE 9.
Indicators of public spending on agriculture (low- and middle-income countries)

REGIONS	PUBLIC SPENDING ON AGRICULTURE PER WORKER (CONSTANT 2005 PURCHASING POWER PARITY DOLLARS)				AOI FOR PUBLIC SPENDING			
	1980–89	1990–99	2000–04	2005–07	1980–89	1990–99	2000–04	2005–07
East Asia and the Pacific	48	69	108	156	0.31	0.48	0.49	0.59
Europe and Central Asia	-	413	559	719	-	0.29	0.35	0.36
Latin America and the Caribbean	337	316	309	341	0.96	0.86	0.56	0.38
Middle East and North Africa	458	534	640	677	0.34	0.37	0.37	0.30
South Asia	46	50	53	79	0.24	0.21	0.21	0.27
Sub-Saharan Africa	152	50	51	45	0.30	0.17	0.14	0.12

SOURCE: FAO, 2012a.
Note: Calculations include 51 low- and middle-income countries for public spending on agriculture per worker and 41 low- and middle-income countries for AOI for public spending. For countries in Europe and Central Asia estimates are from 1995 to 2007. The AOI for public spending equals the agricultural share of government spending divided by the agricultural share of GDP.

Public investment in agriculture is required to foster the public goods that are necessary for generating the rural public capital needed for agricultural production. Evidence shows that investments in public goods have much higher returns than other expenditures, such as general subsidies. However, what constitutes a public good is not always clear-cut and may differ depending on the context. Even though some types of investment are known to bring high economic and social returns, they are not always given the highest priority in budget allocations.

Understanding the impact of different types of public investment and expenditure on agricultural performance and poverty alleviation can help guide public investments towards higher returns. However, this is not an easy task. As noted in *SOFA 2012* (FAO, 2012a), early studies of the impact of aggregate agricultural expenditures on growth and poverty reduction found diverging results. One of the earliest studies in this field (Diakosavvas, 1990) found that government expenditure on agriculture had a strongly positive effect on sector performance. On the other hand, a comparative analysis of data for 100 countries failed to find a statistically significant effect of agricultural spending on growth in per capita GDP (Easterly and Rebelo, 1993).

However, country studies in several regions have found positive relationships between government expenditure on agriculture and growth in agricultural and total GDP. These studies confirmed that the type of expenditure is important. In Rwanda, for example, spending on agricultural research increased agricultural GDP by a factor of 3 to 1. The effects were larger for staples such as maize, cassava, pulses and poultry, than for export crops (Diao et al., 2010). In India, expenditures aimed at improving productivity in livestock had greater returns and were more effective in mitigating poverty than general public investment in agriculture (Dastagiri, 2010).

The substantial literature on public investment in agricultural R&D shows that, over the past 40 years, it has been one of the most effective forms of public investment. Because R&D drives technical change and productivity growth in agriculture, it raises farm incomes and reduces prices for consumers. The benefits multiply throughout the economy, as the extra income is used to purchase other goods and services, which in turn create incomes for their providers. The welfare effects are large and diffuse. Investment in agricultural R&D benefits many people far removed from agriculture. As a result the benefits of this investment are not always recognized as stemming directly from agricultural research (Alston et al., 2000; Fan et al., 2000; Evenson, 2001; Hazell and Haddad, 2001; Fan and Rao, 2003).

A review of 375 applied research programmes and 81 extension programmes (Evenson, 2001) found that in four-fifths of the applied research programmes and three-quarters of the extension programmes the reported rates of return on investment were greater than 20%. In many cases, returns exceeded 40%. Alston et al. (2000) reviewed 292 studies covering the years 1953 to 1997 and found average rates of return on agricultural research of 60% in developing countries. In an update of that study, Alston (2010) found the global rate of return to R&D to have been consistently high.

Recent country level studies support the findings of these comprehensive reviews. For example, research in Thailand is estimated to have a significant positive impact on TFP and a marginal rate of return of 30% (Suphannachart and Warr, 2011). Analysis of an extension service in Uganda reveals rates of return of between 8 and 36% (Benin et al., 2011).

Ensuring enhanced expenditures on agricultural R&D is clearly a priority. IFPRI research shows that the returns on different types of public investment in agriculture change depending on the stage of economic transformation. Governments need to adapt their investment priorities if they are to avoid declining returns per dollar spent, attract more private sector investment along value chains, and enhance the complementarities and synergies between the two (Fan, 2008). This typically means a shift from direct public investment along value chains to a focus on investing in the public goods that are not attractive to the private sector, such as rural roads, electrification and agricultural research.

5.3 INVESTMENT BY THE CORPORATE PRIVATE SECTOR

Private sector investments include those made by households and farmers, and by the domestic and foreign corporate sector. Due to a lack of available data on corporate sector investment in agriculture, case studies from Asia (Indonesia, Lao People's Democratic Republic and Viet Nam) and Africa (Kenya, South Africa, the United Republic of Tanzania and Uganda) were conducted. In the countries studied, there was no available information on the volume of investment in agriculture by the private corporate sector, both domestic and foreign, including multinationals. The information that was compiled showed that the amount invested by the corporate sector was negligible. Most went into high-value commercial crops and the upper end of the value chain (i.e. agro-industries responsible for processing and distribution, and the production of inputs, such as seeds and fertilizers). Very little investment went into the production of the staple crops. There has been, however, phenomenal growth of private sector SMEs, mainly in agroprocessing (FAO, 2012c).

Because the corporate private sector is a high potential source of investment in agriculture, the case studies tried to understand why there is so little corporate sector private investment in agriculture. To do this, systematic interviews with leading private sector investors were conducted and technical consultations were held: one in Bandung, Indonesia in November 2012 for Asian countries; and another in Addis Ababa, Ethiopia in May 2010 for African countries. Executives from private firms, commercial and development banks, investment promotion agencies, and some farmer/agricultural chambers and associations, as well as government policy makers and planners participated in the consultations. The purpose of these initiatives was to:

- Obtain a better understanding of the views and perceptions of corporate private sector investors concerning investment in agriculture, including their perspective on diversifying the agricultural sector.
- Identify the drivers of corporate private investment and the factors that limit their investment.

The perception of all interviewees is that farming presents a challenge for investments. Inadequate profitability was not explicitly expressed as a concern. However, all stressed their perception that the sector is risky. They implied that they cannot make money unless the risk level is moderated. Aspects of risk identified include:

i. The changeability of public policies, procedures and actions is very disruptive. Government authorities carry out erratic, politically motivated policy changes, including frequent interventions in cereal markets, such as the arbitrary setting of prices and bans on trade. The ready availability of food aid and unpredictable imports of commercial foods also disrupts markets for private producers.

ii. The vagaries of the weather and consequent unpredictable production prospects due to limited public investment in irrigation, lack of electricity for irrigation, and unpredictable availability and varying prices of diesel for irrigation.

iii. Land tenure systems in the Eastern African countries do not guarantee long-term rights and can change in response to the whims and resentments of clans or tribal and national leaders.

Most participants in the workshops and the interviewees also indicated that, broadly speaking, the main constraint is the lack of adequate business-friendly socio–economic climate, particularly in the agrofood sector. In their view, this sector is the most politicized, because of concerns for food security and peoples' particular relationship with the land.

5.4 FOREIGN DIRECT INVESTMENT (FDI)

FDI is often referred to as a growing source of finance with major potential for agricultural development and investment. However, because of the lack of comprehensive information, arriving at exact and comprehensive estimates of FDI in agriculture is difficult. Available information on global FDI flows to agriculture is generally incomplete due to the poor reporting, collection and dissemination of data. In addition, the data are sometimes kept secret because of the sensitive nature of the investments. The following analysis uses data from the United Nations Conference on Trade and Development (UNCTAD), whose dataset contains the most comprehensive data on FDI. UNCTAD provides data on FDI in the agriculture sector (agriculture, hunting, forestry and fishing) and the food sector (food, beverages and tobacco).

According to the UNCTAD dataset, FDI inflow to the agriculture sector increased significantly from US$1.9 billion in 2006 to more than US$6 billion in 2007 and 2008. This inflow decreased to US$4.7 billion in 2009, mainly because of the economic crisis. However, even during the 2008 peak, the value (US$6.8 billion) was still significantly smaller than the flow of FDI to the food sector, which stood at US$91.7 billion.[8]

The share of agricultural FDI in total FDI is very low. In 2008, it was 0.7% globally and varied between 0.07% for high-income countries and 2.9% for low-income countries. On the other hand, 9.8% of the global FDI inflow went to the food sector. FDI inflow in the food sector varied between 13% for high-income countries and 1.5 percent for low-income countries (Tables 10 and 11). Except in low-income countries, the share of FDI inflow to the food sector in relation to total FDI is much higher than the share of agricultural FDI.

FDI has played a minor role in the primary agriculture sector compared to the food industry. Given that the value of FDI into the agriculture sector is very small relative to the size of agricultural capital stock, it can be concluded that its contribution to capital formation has been insignificant.

TABLE 10.
FDI inflows to agriculture, hunting, forestry and fishing (by income category in 2008 in current US$ million)

Income category	Country no.	Agriculture FDI	Total FDI	Ag FDI/ total FDI (%)
High	17	377.5	580,099.7	0.07
Upper middle	25	5,561.5	375,247.2	1.48
Lower middle	15	697.0	39,372.4	1.77
Low	10	192.1	6,527.7	2.94
TOTAL	67	6,828.1	1,001,247.0	0.68

SOURCE: FAO, 2013a, and author's calculations.
Note: Computed based on data of 67 countries of which data of FDI inflows to the agriculture sector are available. Total FDI is based on data of the 67 countries.

---

[8] However, it should be noted that it is the end-stage activity that is reported, i.e. if a company invests in land to grow relevant crops, and process and produce biofuel or juice, this would be reported as investment in processing. In this case, it is difficult to assess the trends in broad terms, except with very detailed micro-level data at the firm or enterprise level.

TABLE 11.

FDI inflows to food, beverages and tobacco (by income category in 2008 in current US$ million)

Income category	Country no.	Food FDI	Total FDI	Food FDI/ total FDI (%)
High	21	79,437.3	606,666.3	13.09
Upper middle	13	12,003.3	317,051.6	3.79
Lower middle	5	243.5	8,291.6	2.94
Low	2	28.3	1,901.5	1.49
TOTAL	41	91,712.5	933,910.9	9.82

SOURCE: FAO, 2013a, and author's calculations.

Note: Computed based on data of 41 countries of which data of FDI inflows to the food sector are available. Total FDI is based on data of the 41 countries.

Although FDI has experienced a large surge recently, flows to the agriculture and food sectors are still relatively low compared to other sectors. Within the agriculture and food sectors, FDI is concentrated mainly on the upstream activities (processing, manufacturing, trade and retail). Primary agriculture is left to public sector and the farmers.

FIGURE 2.

Share of agriculture and food in total FDI flows (%)

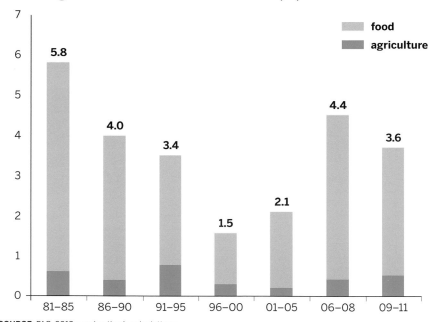

SOURCE: FAO, 2013a, and author's calculations.

Figure 2 depicts the evolution of trends in the share of agriculture and food in total FDI inflows. Despite its importance, since the 1980s, global FDI flows to these sectors have never exceeded 6%. Since the 1980s, the share reached its lowest level (less than 2%) between 1996 and 2000. It has risen since; during 2006–08 the share stood at a modest 4.4% of total FDI flows globally.

As noted earlier, at an aggregate level FDI in agriculture has been relatively small. However, as it has been mentioned previously, agricultural development depends on the simultaneous growth of agricultural production and the value chains to which it is linked. Given that FDI plays a relatively important role in the upper end of the value chain, some selected country case studies were conducted to assess the relative importance of FDI in agriculture. Tables 12, 13 and 14 provide data for Brazil, Cambodia and Thailand. In Cambodia, a low-income country, FDI into agriculture accounts for a larger share than FDI into the food sector. In Thailand and Brazil, which are middle-income countries, FDI into agriculture accounts for only a small share in total FDI and is much smaller than FDI in the food sector.

TABLE 12.
FDI in agriculture in Cambodia (US$ million)

	2000	2005	2006	2007	2008	2009	Annual change 2005–09 (%)
Total FDI	160	684	2,373	1,345	6,866	2,101	32.39
Agriculture (% of total)	1.3 (0.81)	9.1 (1.33)	232 (9.78)	273 (20.30)	74 (1.08)	573 (27.27)	181.69
Food processing (% of total)	-	-	-	10.8 (0.80)	0	0	-

SOURCE: Hang *et al.*, 2011.

TABLE 13.
FDI in agriculture in Thailand (Baht million)

	2000	2005	2006	2007	2008	Annual change 2005–08 (%)
Total FDI	256,282	740,717	1,274,047	857,203	697,567	-1.98
Agriculture (% of total)	34 (0.01)	686 (0.09)	387 (0.03)	252 (0.03)	606 (0.09)	-4.05
Food processing (% of total)	4,288 (1.67)	8,484 (1.15)	18,571 (1.46)	17,336 (2.02)	18,432 (2.64)	29.52

SOURCE: Waleerat and Nipawan, 2011.

TABLE 14.
FDI in agriculture in Brazil (US$ million)

	2003	2004	2005	2006	2007	2008	Annual change 2005–08 (%)
Total FDI	10,100	18,100	15,100	18,800	34,600	45,100	44.01
Agriculture, livestock (% of total)	170.5 (1.69)	166.3 (0.92)	210.2 (1.39)	176.1 (0.94)	316.9 (0.92)	498.1 (1.10)	33.32
Food and beverage (% of total)	409.4 (4.05)	5,345.5 (29.53)	2,074.8 (13.74)	739.3 (3.93)	1,816.7 (5.25)	2,238.2 (4.96)	2.56

SOURCE: Central Bank of Brazil.

5.5 OFFICIAL DEVELOPMENT ASSISTANCE (ODA)

ODA is a significant component of overall resources for most developing countries. By and large, ODA is committed through the public sector[9] and could be counted as public expenditure. In view of this, it is assumed that 40–50% of ODA is investment for capital formation. Using the FAO external assistance to agriculture (EAA) dataset and assuming that ODA makes 40% contribution to capital formation, the figures in Table 15 present ODA as a percentage of total investment for ACS. As the results show, ODA's contribution to capital formation is very small across all the regions. Even if it is assumed that ODA makes a 50% contribution to capital formulation, the difference in ODA's relative contribution to total investment is negligible.

TABLE 15.
Percentage of ODA in farm-level ACS

Region/Country groupings	2005	2006	2007
East Asia and Pacific (9)	0.5	0.2	0.6
Europe and Central Asia (9)	0	0	0
Latin America and Caribbean (11)	0.4	0.4	0.4
Middle East & North Africa (7)	0.9	0.9	0.9
South Asia (7)	0.8	0.5	1.3
Sub-Saharan Africa (11)	1.7	1.7	3.2

SOURCE: FAO EAA dataset.

---

[9] Some ODAs are also committed through civil society organizations (CSOs) and non-governmental organizations (NGOs). However, there are no available data on how much is channelled through the public sector and how much through the NGOs and CSOs.

Drivers of investment in agriculture for increased production and productivity

From the preceding sections it can be seen that the main source of investment for agriculture for farm-level capital formation is farmers themselves. The public sector comes next at a distant second. In terms of promoting investment by farmers and the private sector in general, there has been much discussion and debate over the complementarity between the two and the importance of appropriate policies for creating an enabling environment for investment. In the literature it is often argued that public sector investment and an enabling environment promotes private investment. For this reason the policy focus has heavily tilted in favour of this approach.

6.1   PUBLIC–PRIVATE COMPLEMENTARITY IN INVESTMENT

The relationship between public and private sector investment remains a matter of great debate. In India, the focus on the subject became intense due to its implications for policy directions influencing capital formation in Indian agriculture. There were several empirical studies conducted (Rath, 1989; Gandhi, 1990; Shetty, 1990, 1996; Storm, 1993; Rao and Gulati, 1994; Mishra and Chand, 1995; Dhawan, 1996,1998; Mishra, 1996; Mishra and Hazell, 1996; Mitra, 1996; Rao, 1997; Misra, 1998; Chand, 2000; Roy, 2001; Gulati and Bathla, 2002; Chand and Kumar, 2004). However, the debate on the impact of public investment on private investment remains inconclusive. Depending on the dataset, time frame and model specifications there are three conclusions that have been reached regarding whether there is complementarity or not between public and private investment from the above studies. Some believe public investment induces private investment, which would suggest a strong complementarity between these two types of investment. Others see only weak complementarity between these two types of investment, while others see no apparent complementarity at all.

Without ignoring other determinants of private investment, it does appear that there is more evidence to confirm the complementarity between public and private investment.[10] However, the degree of elasticity of private investment with public investment varies depending on the period of study; the choice of variables on public sector investment and their measurement; the construction of appropriate lag time for determining impact; and the specification of structural equations. It is also argued that complementarity depends on the kinds of public investments that are made. Many public investments in agricultural R&D, rural roads, electrification and education give high returns in terms of agricultural growth and poverty reduction and are highly complementary to private sector investments (Fan, 2008). Figure 3 provides evidence of a positive relationship between public expenditure in agriculture per worker and the growth of on-farm capital stock per worker. Public expenditure on agriculture seems to also support the formation of on-farm capital stock. However, the large variation of observations around the fitted trend line indicates that other factors affect the relationship. These factors likely include the composition of public expenditures on agriculture and their effectiveness.

---

[10] All the case studies have reported a positive inducement/complementary effect of public investment on both corporate and farm household investment. But in the Plurinational State of Bolivia, public investment has more inducement effect on corporate/agro-industry investment. However, it is not clear whether public investment pattern has been designed more to support the corporate sector.

FIGURE 3.
Agricultural expenditure and growth in ACS per worker, 1990–2007

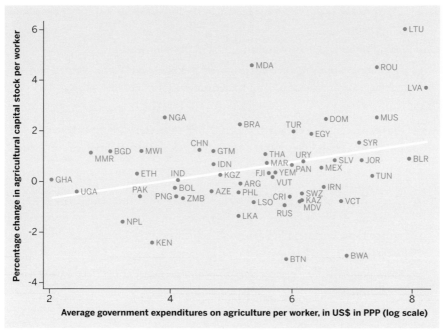

SOURCE: FAO, 2012a.

The challenge lies in gaining a better understanding of the factors that foster private farm household investment in agriculture and identifying priorities for public investment in and for agriculture. To meet this challenge, each country's stage of agriculture growth and development and broader national development objectives must be kept in mind.

6.2 POLICIES AND THE ENABLING ENVIRONMENT FOR INVESTMENT

Creating a favourable policy environment is considered to be an effective way to promote private investment in agriculture. Considerable attention has been paid to analysing what constitutes an 'enabling environment' and what are the factors that would contribute to its creation. By and large, the provision of an enabling environment lies with the government. According to the World Bank (World Bank, 2004), the roles of government in providing a good general investment climate include:

- Ensuring stability and security, including the safeguarding of rights to land and other property, contract enforcement and crime reduction.
- Improving regulations and taxation, both domestically and for international investments.
- Providing infrastructure and financial market institutions.
- Facilitating labour markets by fostering a skilled workforce, crafting flexible and fair labour regulation and helping workers cope with change.

Some of these roles go beyond what are termed as 'policies'. Policies provide and/or create the enabling environment for people to invest. Policies can do this in two ways:

- Through specific interventions to stimulate investments, such as interest subsidies and correcting rural finance deficiencies (e.g. asymmetric information, high transactions costs, no insurance schemes, no loan-resale markets).
- Through broader price-related policies that affect the investment environment.

The creation of an enabling environment goes well beyond policies and is not exclusively for investment in agriculture. They would apply equally for investments in all sectors, including agriculture (FAO, 2012a), as well as for overall quality of life.

In recognition of the importance of an enabling environment for agricultural investment, the OECD and the New Partnership for Africa's Development (NEPAD) have developed a comprehensive draft policy framework for promoting investment in the sector (OECD, 2011). The framework emphazises the complexity and diversity of the issues involved in ensuring an appropriate environment for agricultural investment and the extent to which the required policies and institutions transcend agriculture (Annex 2).

6.3 POLICIES, TRENDS AND INCENTIVES TO INVEST

The relevant policies for promoting investment are those that drive a wedge between value added as observed and the value added that would otherwise prevail in an undistorted opportunity costs situation with the use of border prices.

These policies include:

- Tariffs and export taxes, trade quotas, price subsidies or taxes on outputs and intermediate inputs, domestic price interventions, credit subsidies or rationing.
- High protection to non-agriculture sector and/or selected industries.
- Exchange rate misalignment.

In the post-war period, many developing countries enacted policies that were biased against agriculture and created disincentives for investment and production. These policy-induced 'distortions' to agricultural incentives in developing countries and their associated costs for 18 countries were documented by Krueger *et al.* (1988, 1991). Later, structural adjustment programmes led by the World Bank and the International Monetary Fund (IMF), focused on reducing the policy-induced distortions. As a result, over time, policy reforms have changed the levels of protection and taxation of agriculture in many countries. These broad trends for 75 developed and developing countries are described in a recent study by the World Bank (Anderson and Valenzuela, 2008; Anderson, 2009).

The relative rate of assistance (RRA), an indicator of policy-induced price distortions to agriculture, measures the extent to which government policies affect farm prices relative to other sectors. The RRA also provides an indication of the degree to which a country's overall policy regime is biased for or against agriculture. A positive RRA implies that agriculture is favoured or subsidized relative to other sectors, while a negative RRA indicates that agriculture is penalized or taxed (Anderson and Valenzuela, 2008). The analysis of average RRAs over time shows the dramatic differences between developed and developing regions regarding policy positions towards agriculture. From the mid-1950s, agriculture was taxed heavily in many of the low- and middle-income countries of Asia, Latin America and sub-Saharan Africa, while the sector was increasingly protected in many high-income countries.

Beginning at different times, the low- and middle-income countries have gradually reduced the bias against agriculture and the high-income countries have reduced the degree of support (Fig. 4). For the low- and middle-income countries, the movement of the average RRA towards zero is due to a decline both in the taxation of agriculture and in the protection of other sectors. The extent and speed of the changes vary across regions.

FIGURE 4.
Relative rate of assistance to agriculture, by region, 1955–2007

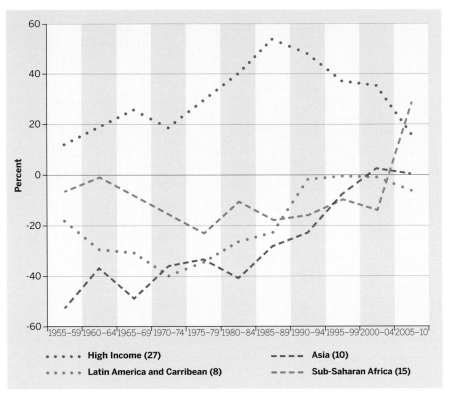

SOURCE: FAO, 2012a.
Note: Five-year weighted averages based on agricultural production valued at undistorted prices. The number of countries included in each group is shown in parentheses.

The impact of policy-induced distortions on levels of agricultural investment in low- and middle-income countries has reduced over time. However, the convergences of averages hide considerable variation within regions and income groups. Policy-induced price distortions remain significant in many countries. In general, low- and middle-income countries are still more likely to tax agriculture whereas high-income countries subsidize it. Distortions in some countries are quite extreme. This suggests that efforts to increase investment in agriculture in low- and middle-incomes are still being hindered by policy-induced distortions at both ends of the spectrum.

Reducing the remaining price-distorting policies would improve incentives to invest. It would also lead to better resource allocation by directing investment toward the activities and industries in which each country has its strongest comparative advantage. In the case of countries that still discriminate against agriculture, it is likely that such reforms would boost investment in agriculture, especially areas and subsectors with the highest payoffs. Several studies have estimated the impact of distorted incentives to agriculture on national and global economic welfare, economic growth and poverty.

Anderson *et al.* (2009) provide a combined retrospective and prospective assessment using an economy-wide modelling exercise. They use the World Bank's Global Linkage Model (van der Mensbrugghe, 2005) to quantify the impacts both of past reforms (up to 2004) and of potential benefits from removing remaining distortions. Their results confirm the significant gains to agriculture especially in developing countries that could be made by removing distortions to price incentives (FAO, 2012a).

The dynamic effects of price distortions are analysed by Anderson and Brückner (2011). They examine the effect of moving the RRA towards zero on overall economic growth of sub-Saharan African countries. Given that most countries in the region currently tax agriculture, removing these price distortions would have a significant positive effect on their overall economic growth rate. These results show that taxing agriculture relative to other sectors reduces national economic welfare and reduces overall output growth over time (FAO, 2012a).

According to FAO (FAO, 2012a), a favourable investment climate is indispensable for investments in agriculture. However, it is not sufficient to allow many smallholder farmers to invest and ensure that large-scale investment meets socially desirable goals. Also, some scholars argue that price incentives and favourable terms of trade alone would not bring about significant improvements in investment in agriculture because aggregate agricultural supply is inelastic (e.g. Mellor–Delgado, mid-1980s). They argue that structural changes, specifically public inputs such as research, infrastructure and input delivery systems, as well as cost reduction through publicly provided inputs, are the essential ingredients for agricultural growth.

The removal of the policy distortions, though necessary for a favourable enabling environment, will not be sufficient to ensure that private investment will take place, particularly in those countries where farm-level agricultural capital stock per agricultural worker is low. Agriculture in these countries, mainly in South Asia and sub-Saharan Africa, is largely dominated by small and marginal farmers who often lack the capacity to respond to incentives. As noted earlier, the challenge lies in gaining a better understanding of the factors that induce private farm households to invest in agriculture.

6.4 DRIVERS OF HOUSEHOLD INVESTMENT IN AGRICULTURE

In the agriculture sector of developing and developed countries alike, the dominant mode of production is the 'family farm', where the family owns, manages and provides the primary source of labour (Brookfield, 2008). It is their investment that drives agricultural production and, as noted earlier, they are the main source of investment for farm-level capital formation. It is crucial to understand their investment behaviour and the kinds of policies that can enhance it.

As households, family farms have a number of investment options besides their farms. Their investments are often intended to accumulate the following five types of capital (Benoit-Cattin, 2011; Bisaliah *et al.*, 2013):

- Human capital, consisting of the household members and their education, experience, health and nutrition.
- Natural capital, consisting of common property resources supporting agricultural activities (e.g. grasslands, forest, water) and private resources such as farm land, animals and trees.
- Physical and fixed capital, consisting of production assets like tools, machines, building, transport vehicles and information technologies.
- Financial capital, consisting of savings and access to credit.
- Social capital, consisting of family networks, professional (e.g. credit or marketing groups) and non-professional (religious, veterans) organizations, and political affiliations.

Some of these different types of capital are highly complementary for increasing farm productivity. For example, investments in human capital improve the productivity and skills of family members working on the farm. These kinds of investment are typically not included in standard measures of farm capital formation. However, they are very important in leveraging the returns to the more standard measures of farm capital formation, which cover only natural and physical capital. Investments in social capital can also enhance the management of common property resources and improve access to markets and credit, all of which can contribute to greater productivity of the farm.

On the other hand, some types of capital have little to do with farm productivity. Investments in human and social capital can increase options for sharing risk (e.g. within local kin networks) or facilitate migration and non-farm employment (e.g. investments in the education of children or maintaining links with kin in urban areas). Farm families must invest labour, time and financial resources to build up these different types of capital. They can be expected to invest more in the types of capital that yield the best returns. However, the important point to note is that their 'returns' needs to be measured in several dimensions, including increased food production for household consumption, a larger farm and farm house for living and as a fixed asset, social status, more cash income, greater security, or additional livelihood options.

Case studies were conducted in Bangladesh, Brazil, Burkina Faso, China, India, Nepal, Egypt, Ethiopia, Malawi, Mali, the Plurinational State of Bolivia, the Republic of Korea, South Africa and Zambia to find out what drives investment for farm-level capital formation. The drivers of farm-level capital formation, as found in a review of the case studies, are presented in the following pages.

Household savings

What emerges from the studies is that the farmers who cannot save, cannot invest. If they cannot save, not only can they not invest, they also cannot access credit. Even when they do get credit, they use the loans for immediate consumption and become more and more indebted. Saving is also needed to generate assets (again through investment), which gives access to credit. In fact, in a situation of severe credit constraint, the saving and investment decision of an agricultural household can hardly be separated since its investment decisions are linked to its saving decisions.

In a survey of 51,770 households spread over 6,638 villages across India, it was found that the estimated annual savings of all farming households during 2002–03 was negative, at 69,348 crore Indian rupees. As a result, the ratio of farm sector savings to overall GDP was estimated at -2.8% for the year 2002–03. This has led to increased indebtedness and a decline in farm-level capital formation in rural areas. Taking into account the indebtedness of farming households from the All India Debt and Investment Survey 2003, the proportion of cash loans as proportion of overall GDP turns out to be 3.3% during 2002–03. Interestingly, this ratio is quite close to the amount of dissavings, indicating that the gap between income and consumption expenditure is financed by borrowings.[11]

Survey results also show that less than 1% of farmers in Zambia and less than 2% of the rural population in Nigeria have access to formal credit (Meyer, 2011). In India, 45% of smallholder farmers do not have a formal savings account, and 69% do not have access to a formal credit account. Nearly 40% of farmers in Honduras, Nicaragua and Peru are credit-constrained (AgriFin,

---

[11] Report of the High Level Committee on Estimation of Saving and Investment, Ministry of Statistics and Programme Implementation, Government of India, New Delhi, 2009.

2010). At regional levels, the share of adults who do not have access to formal finance institutions is 65% in Latin America, 80% in sub-Saharan Africa, and 58% in South Asia and East Asia (Chaia *et al.*, 2009).

Household farm savings have long been recognized as the most important source of on-farm investment (Wilcox, 1943). Recent data show the continuing importance of household savings for financing investments by rural entrepreneurs, including smallholder farmers (Collins *et al.*, 2009). However, the determinants of household savings are complex and not fully understood. They are affected by income levels, property rights, the presence or absence of social insurance systems and demographics. There may also be cultural factors that affect the propensity to save. Due to the fact that saving is influenced by a number of socio-economic factors and cultural and psychological variables, the available evidence and estimates of savings functions, either at aggregative or disaggregate level, is not satisfactorily conclusive. Most analytical work has focused on one or a few explanatory variables at a time. Among these most notable variables are: income in its different forms, such as absolute income, relative income and permanent income; wealth; land tenure; external resources; rural credit; taxation; demographic characteristics; rural market structure; price level; production possibilities; and technology.

The impact of real interest rates on household savings is uncertain in the theoretical and empirical literature. Warman and Thirlwall (1994) using data from Mexico for 1960–90, observed a strong positive effect of interest rates on financial savings, whereas their effect on total savings and private savings is insignificant. Masson *et al.* (1998) have found that real interest rates have a positive and significant effect on private savings in industrial countries and a negative insignificant coefficient for developing countries. Loayza *et al.* (2000) found insignificant and negative coefficients for real interest rate changes on private savings, and the results were not robust across samples. In India, most of the studies (Muhleisen, 1997; Loayza and Shankar, 2000; Athukorala and Sen, 2001, 2003) have taken proxy variables, such as bank branch expansion and total institutional lending to the private sector, to represent financial sector liberalization, in order to examine their impact on private savings and obtained insignificant impact of these variables.

However, notwithstanding the multiplicity of factors affecting savings, it can be argued that savings behaviour is largely influenced or determined by factors related to ownership, transferability and transformability (Alamgir, 1976).

Ownership of savings refers to the question of total control of the amount saved in physical, legal and moral terms. Transferability of savings refers to the extent to which it is possible to transfer savings from one person to another; transfer savings in one form of asset into another; transfer savings from one sector of the economy to another; and transfer savings from one set of economic units to another within the same sector or outside it. Transformability on the other hand relates savings to investment and the growth process. The importance of saving lies in the fact that it helps maintain and improve existing productive capacity. It is absolutely essential that saving be related directly or indirectly to creation of productive capacity. The crucial question related to any particular form of saving is to what extent it is transformable into productive capacity.

Income and remittances

Closely associated with savings is income. Remittances have recently become a main source of rural household income in many developing countries. They were found to be important source of investment in agriculture for the development of family farming and particularly for making the shift from subsistence agriculture to market-oriented production. Migration is predominantly a family decision. It is the family that decides whom to send, mobilizes the cost of migration and, in return, receives remittances for the wider benefit of the family. However, it should be noted that large part of remittances are used for immediate consumption, health and education. Only a small proportion, around 10–12%, is invested in agriculture (FAO, 2012a). In Bangladesh, the average annual household saving for migrants is 87,583 Bangladeshi Taka (BDT) compared to 28,957 BDT for non-migrant households. Migrant households save about three times more than a non-migrant household. However, in terms of percentage of income, a migrant household saves 25% of their total annual income, whereas non-migrant households save about 22% of their total annual income. There is a large difference in the amount saved but the difference between the proportion of income saved is small (Hossain, 2012).

The impact of remittances on agriculture and smallholder farmers depends on the particular context. For example, in some rural areas of Morocco, emigration causes agricultural production to fall in the short term because there is less farm labour available, but the long-term effects are positive as remittances are invested in agriculture (de Haas, 2010). Similar results have been found in five southern African countries. Domestic crop production falls initially, but in the longer term, crop productivity and cattle ownership are boosted by remittances and higher domestic plantation wages (Lucas, 1987). In Ghana, the initial negative impacts of migration were compensated over time by remittances, which stimulated both farm and non-farm production (Tsegai, 2004).

Evidence from Asia also shows positive long-term effects of remittances. In the Philippines, Gonzalez-Velosa (2011) found that remittances were invested in working capital and served as insurance. Farmers who received remittances were more likely to grow high-value crops, adopt hand tractors and threshers, and invest in irrigation. There was no negative impact on production as there were no labour constraints. Overall, they found that remittances facilitated agricultural development. In Bangladesh, Sen (2003) and Kazi (2012) found evidence that off-farm labour, including migration, in combination with other diversification strategies, has allowed poor rural households to accumulate assets, purchase farm machinery, agricultural land, tractors and tube wells, and make use of more human capital. Mendola (2008) shows that Bangladeshi farmers with an international migrant in the family are more likely to adopt rice varieties with greater yield variability. Evidence from India also supports the argument that when farming is profitable, agriculture attracts remittances for investment. For example, Oberai and Singh (1983) found that in Punjab, a fertile area of India, remittances were invested in agriculture.

FIGURE 5.
Cash farm income and farmer's expenditures for specified capital goods, 1910–39

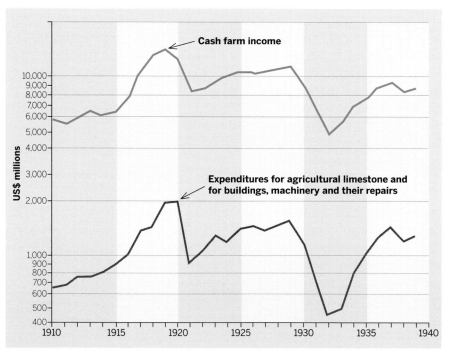

SOURCE: Wilcox, 1943.

The positive relationship between farmer's income and investment, (i.e. purchases of capital goods) was observed in the USA as early as 1910. Wilcox (1943) noted that the farmers' purchases of capital goods since 1910 show quite clearly (as would be expected) that investment in capital in agriculture is closely related to current income rather than to variations in interest rates, prospective income, or dates when improved machines were perfected. The close relation between variations in farm incomes and farmers' investments in capital goods is shown in Fig. 5.

This same correlation exists between variations among farmers in their investment in working capital over a period of years and the size of their average annual net farm income. Only farm families with the relatively high incomes make capital investments on the marginal principle in order to maximize their net income over a period of years.

Property rights

The property rights that farmers have over their land was found to be important in determining whether they are willing to make investments that improve the long-term productivity of local resources (e.g. trees, contouring, irrigation). Secure property rights are also important for obtaining credit needed to make long-term investments.

In many cases, it was found that it is not only property rights per se that are critically important, but also the ability to uphold those rights (i.e. the cost and speed of accessing legal redress in case of disputes and perception of fairness of the administrative and judiciary system). In many countries, it was found that many farmers are involved in court cases dealing with land for long periods of time. In fact, most of the legal cases in rural areas are related to disputes over land. Therefore, farmers are reluctant to invest in disputed land.

There is a considerable literature demonstrating that tenant farmers typically invest less in long-term capital formation than farmers who own their land. However, even among farm owners, the type of property rights they have can be important. Property rights vary widely across developing countries. This is primarily because developing countries are at various stages in the transition from indigenous, community-based tenure systems to registered and legally recognized, private property arrangements (Deininger, 2003). The vast majority of African farmers, for example, continue to hold and use their land under customary tenure arrangements embedded in social relations that define the rights and obligations between individuals and groups. These rights and obligations pertaining to land are not recorded and do not require any documentation because they are guaranteed by the collective action of their local communities. A World Bank estimate indicates that in Africa only 2–10% of land is registered under formal tenure (Deininger, 2003). These customary land tenure arrangements have served Africa's farmers well in the past. They have provided secure use rights to individual farmers while retaining sufficient flexibility to recognize multiple user rights over some resources (e.g. shared tree rights and pastoralists' grazing rights on cropped land after harvest) and to protect the access rights to some land for qualified members of the community or tribe.

There is evidence that customary land rights systems evolve over time towards more individualized private rights in response to economic and demographic pressures (Migot-Adholla *et al.*, 1991). However, these changes take time and rarely happen within the time frame of a major investment or technological change. The lack of codified land rights and a legal title can also expose small farmers to land grabs by the rich and powerful. This has happened often in the past in the context of large-scale irrigation or infrastructure projects that rapidly raised land values. Existing land rights systems also discriminate against women in many societies, making it difficult for them to control or inherit land even when they are the primary farmer. Widows are especially vulnerable to losing established rights to use the land. By restricting land transfers to members of the same community or tribe, the indigenous tenure systems can also hinder the development of efficient land markets for sale or rent.

Farm size
Although studies have found that small farmers are nearly, if not more, productive than the big ones, the case studies typically found that larger farms innovate and invest more than small farms. This was also observed in India by the All India Survey of Savings and Investment (High Level Committee on Estimation of Saving and Investment: Ministry of Statistics and Programme Implementation, Government of India, New Delhi, 2009). Part of this may reflect the greater dependence of large farms on hired labour, which gives them greater incentive to adopt labour-saving technologies, such as machines and herbicides. The ability of larger farms to innovate and invest more may also reflect their wealth, opportunities for higher savings and privileged access to credit and markets.

Human capital
The levels of education and literacy of the heads of households are widely associated with greater innovation and investment in farming, as is farmer training and extension. Gender is also important, but often in a perverse way, as women farmers are often discriminated against when it comes to gaining access to new knowledge, technologies, property rights and credit.

Available public infrastructure

Access to infrastructure, such as roads, electricity, information technology and irrigation schemes, can all have an important bearing on farmers' willingness to invest. Public investment in rural infrastructure is highly complementary to farmers' own investments. Likewise, private agro-industrial investments that increase farmers' access to modern inputs and strengthen their links to markets for selling products and obtaining added value from processing, were found to create additional incentives for greater on-farm investment. Lack of public expenditure in public infrastructure was found to be a major disincentive for farm-level investment in Nepal and the Plurinational State of Bolivia. In Mali and Burkina Faso, public expenditures gave a major boost to farm-level investment, which demonstrates a strong complementarity between public and private investment.

Collective action

Some investments need to be undertaken at the community level and require collective action. Examples include investments in common property resources, such as water or grazing lands, or the organization of neighbouring farmers to undertake land terracing or water storage. Organizing farmers into effective and stable groups for collective action is difficult and time-consuming. Its success depends on a range of physical, social and institutional factors (Ostrom, 1990). Success is also determined by the security of the property rights that the community has over resources.

In recent years, there has been an explosion, driven largely by NGOs and some donors, of community-based organizations engaged in natural resource management. To effectively manage natural resources collectively, community orgainzations need to include all key stakeholders (Uphoff, 2001). In some cases, as for example contouring part of the landscape, only a small group of farmers may be involved. In other cases, as in watershed development projects, collective action may need to involve the whole village. Other villages may need to be involved when the the collective action addresses the management of open rangelands shared with other communities or distant tribes.

Community approaches require effective local organizations to plan and implement agreed interventions. To avoid the monopolization of authority by an elite group and resolve local disputes, broad representation is needed in the governance of local organizations. The start-up and early development of these organizations is usually difficult because communities often lack leadership and technical and administrative skills. Technical training and leadership support from outside agencies (NGOs) has often proved crucial in the early stages of organizational development.

6.5 DRIVERS OF INVESTMENT IN AGRO-INDUSTRIES

The investment decision

The commercial strategies or business models adopted by agro-industrialists for making investment decisions vary widely depending on specific circumstances. However, the investment process is more or less the same in each case.

The basic trading ambition of agro-industrialists can be characterized as that of being able to sell low-cost products into high-value markets, and thereby maximize profit margins. In middle-income countries with less efficient domestic markets, this is likely to lead to an emphasis on export markets, particularly for larger agro-industrial enterprises. The same effect also tends to apply to small countries. For agro-industries downstream from on-farm production, the low-cost objective is more likely to be achieved by sourcing raw material from regions where domestic resource costs (DRCs), mainly land and labour, are relatively low, but agricultural productivity is, or can be made, relatively high. This combination is likely to result in relatively low procurement prices for farm products. However, the likely relative underdevelopment of infrastructure and services needed to make farm procurement possible on a relatively large scale (particularly from smallholder farmers) in low DRC rural economies, offsets the advantage of low procurement prices, as additional investment by agro-industry itself will be required.

The low-cost objective will also drive agro-industry towards larger-scale operations. A corresponding 'scaling up' of small farm production through group action by smallholder farmers, through producer associations or cooperatives will be attractive to agro-industry because it facilitates the farmers' ability to reliably provide the quantities of raw material that agro-industry requires.

Agro-industrial enterprises can achieve high-value market objectives by accessing niche markets in which premium prices are paid for products with special attributes, or the prices available are high relative to the supply cost of the product. Examples of niche market products are organically grown products, such as organic rice from Cambodia; products branded on the basis of their geographical origin such as Blue Mountain coffee from Jamaica; and 'ethical' products such as those traded under the Fair Trade label.[12] A globally significant example of relative value advantage is the trade in fresh fruits, vegetables and cut flowers, such as Peruvian asparagus, Thai baby corn and Ethiopian roses, which are transported by air to industrialized countries.

---

[12] For a time, following market liberalization in the Former Soviet Union, an agro-enterprise in the Krasnodar region of southern Russia sought to develop a market in 'magnetic' sugar.

Investment financing and risk

According to UNCTAD, most of the investments made by agro-industrial companies in agricultural value chains is financed from the companies' own savings and retained profits. Empirical evidence, based on cross-country averages for more than 32,000 firms of various sizes from 100 developed, developing and transition economies for the period 2002–2006 (Annex 1), shows that globally firms finance about two-thirds of their investments from retained earnings and another 16 to 23%, depending on the size of the firm, from bank loans (UNCTAD, 2008). With regard to credit access, the larger corporate players in agro-industry are unlikely to face special problems in raising finance for new investment. As da Silva *et al.* (2009) note, it is agro-industrial SMEs that are particularly liable to be caught in a credit 'trap'; too small to access formal capital markets and too large to rely on family savings or informal lenders. Lack of access to formal capital markets is a reflection of the commonly held view in the banking sectors of many developing countries that agro-industrial SMEs are high-risk and small-return businesses. In Cambodia, for example, the International Finance Corporation (IFC) and the World Bank have recently established a risk-sharing facility that partially guarantees a portfolio of new commercial bank loans to agribusiness (mostly SMEs) in an effort to accelerate lending to a sector traditionally perceived in Cambodia as high risk (World Bank, 2010). Another example from Cambodia is the country's only organic cashew nut business, which, despite having a sound business plan and certified farmers already signed up as potential suppliers, is having difficulty attracting equity investment as of mid-2011. This commonly held view of the riskiness of agro-industrial investment is self-fulfilling; it forces SMEs to focus on limiting their exposure to business risk in input and output markets by keeping the business in the family (i.e. rather than partnering with non-family investors). This strategy means that these businesses will tend to remain small, inefficient, high-cost and poorly informed.

Value-chain partners

Given the significant degree of non-controllability in agriculture-based activity, it is particularly important to agro-industrial enterprises that their upstream and downstream partners in agricultural value chains are reliable and trustworthy. It is often logistically more difficult and costly for agro-processors or agro-distributors to achieve effective vertical coordination of the value chain if upstream supplies have to be sourced from large numbers of small farms rather than a small numbers of large farms. However, this is not always the case; some forms of smallholder agriculture may be better at meeting agro-industrial enterprises' requirements in terms of quality assurance, quantity and delivery schedules. This is especially true where intensive management supervision is essential to ensure high quality (e.g. in horticulture). Much depends on local agrarian institutions and the type of product. In many countries, there are long-established agro-industries, particularly those based on 'plantation' crops (e.g. coffee, cocoa, cotton, oil palm, rubber, tobacco, sisal, sugarcane, tea), which are supplied with raw materials entirely, or at least to a significant extent, by small farmers. However, other things being equal, corporate agribusinesses claim to prefer dealing with a few large farms than with several small farms.

The reality, though, is a little more complicated. As Swinnen and Maertens (2007) point out, although dealing with larger farmers is indeed the preference expressed by companies, in practice they contract with many more small farms than expected. This is consistent with the conclusion of Barrett *et al.* (2012) concerning contract farming arrangements: there is little empirical evidence of a positive or negative correlation between farm size and value-chain participation by farmers. However, it is not clear whether this also applies to farms owned or operated by farmers who belong to some form of farmer organization, such as an agricultural producer association or cooperative, which then becomes the companies' primary point of engagement with the small-farm sector. Many agro-industrial SMEs have little choice in the matter and are more or less obliged to engage with small farmers. A table summarizing the arguments, from the agro-industrial perspective, for and against procuring from small-scale producers is given in Vorley *et al.* (2009).

Generally, the greater the disparity in economic power between value-chain partners, the more difficult it is to build up trust between them. From the small farmers' perspective, there are both advantages and disadvantages of engaging with corporate buyers in formalized arrangements regulated by contract. As already noted, the advantages can include guaranteed access to credit; more advanced agricultural technology with advice and training on how to use it; good quality inputs; and a stable market. On the other hand, meeting the requirements of corporate buyers can mean capital outlays on specialized equipment for a specific production process or particular facilities to ensure compliance with value-chain certification, a higher level of indebtedness, and less decision-making autonomy in production and marketing.[13] The advantages of a guaranteed market outlet may be offset by a contractual price that is lower than prices perceived by farmers to be available outside the corporate market.[14]

---

[13] Indeed, Reardon and Barrett (2000) go so far as to conclude that 'contracts are not the institutional panacea for small farmer involvement in agro-industrialization'.

[14] It is not uncommon, however, in such circumstances for farmers to overlook the fact that high prices in local traditional markets may reflect relatively low-traded volumes, and prices could collapse if the bulk of the corporately contracted supply was diverted into these markets.

This accounts for the finding by Barrett *et al.* (2012) that membership of a cooperative or some other farmer organization matters to small farmers. This membership is important partly because it lowers transaction costs and helps attract contract offers from companies, but also because the contract terms available through farmer organizations are usually better than those available to farmers acting individually.

For the SMEs, particularly the smaller enterprises, value chains are more likely to be personalized with high levels of trust and familiarity considered important. Larger enterprises, for which value chains are relatively non-personalized, will generally look for opportunities that will give them:

i. A high degree of control over total value-chain coordination to minimize supply risk (quantity, reliability, quality, traceability) and where relevant, supply risks on both sides of their position in the chain, as with agricultural processors who must source farm inputs and market outputs.

ii. A sufficiently large exploitable supply base upstream and a sufficiently large potential market downstream (as applicable) to reach the minimum scale of operation required for acceptable cost-efficiency.

iii. Controllable, or at least, predictable margins, with the ability to pass back upstream price changes to downstream closer to the final market (the less competition the better from the company's perspective, as is little or no official regulation of prices or margins).

iv. Economic space in which to grow, whether because market demand is likely to remain unsatisfied; there are prospects of accessing niche markets that pay a premium price; or there is the potential for reducing unit production costs through efficiency improvements that enhance the company's competitiveness.

These preferences tends to push the growth and evolution of value chains in certain directions that are not likely to favour greater participation by small farms or provide the most favourable terms on which they can participate. However, they do not necessarily prevent either of these things.

Importance of location

In localities served by relatively good infrastructure and services, and populated by well organized and trained farmers, many companies are prepared to accept risks and manage them as a matter of routine. Indeed, even where the alternative may be possible (i.e. sourcing from the company's own farms), the additional costs of developing these options may be unattractive.

The benefits, costs and risk calculations change dramatically, however, when the enterprise is a new venture located in a remote and underdeveloped area.[15] The start-up investment is likely to be very high, with the agro-industrial enterprise having to provide by itself almost everything it needs in terms of infrastructure and services. This creates large and costly inventory holdings in order to overcome basic logistical problems caused by weak supply lines. The business will also probably have to pay premium wages to attract and retain skilled staff. On top of this, it will need to train and support local farmers, at an economic cost, so that they can become reliable suppliers. Expectations, among the farmers and their local government organizations, may be high. So may be the scope for misunderstandings as to what each party has agreed to put into the venture in terms of land, labour, technology and capital. Several agribusinesses have failed because of a breakdown in trust with the local farming community, and the consequent withdrawal of local political support for the venture.[16]

All this presents a daunting challenge to agro-industry. Many enterprises in such circumstances opt first to develop their own farms if they can (and solve the problems involved) before engaging with small farmers on a commercial, as opposed to a pilot, scale. Either that or they choose not to locate in the area at all.

Importance of farmer organizations

There is much evidence to show farmers seek out memberships in farmer organizations, such as producer associations and cooperatives. Barrett *et al.* (2012) find that such membership not only lowers transaction costs and helps attract contract offers from agro-industrial companies but also delivers contract terms to farmers that are usually better than those they would be offered if they acted individually. For the agro-industrial enterprise, the costs and risks of relying on large numbers of unorganized small farms to supply the greater proportion of corporate demand may be high.[17] The existence of producer associations or cooperatives of small farmers will generally induce the enterprises to engage with the small-farm sector, even if it entails the possible disadvantage of having to deal from a weaker negotiating position.

---

[15] Barrett *et al.* (2012) point out that the geographic placement effect of corporate decisions to contract with smallholders is commonly overlooked in the literature, p. 72.

[16] In one case in Senegal, for example, the deal was for the company to provide materials and equipment (Jatropha seedlings, drip irrigation, organic fertilizer, termiticide, technical expertise, management, etc.) and for the villagers to provide unpaid labour on the land agreed to be made available by the local community under the auspices of the *Conseil General*. The harvest would belong to the local people and the company would buy it from them. However, after only 18 months or so, local disenchantment with the project started to set in, and people began insisting on being paid for working on the land.

[17] The exception is when the processor/trader has a monopoly over the agricultural product concerned and is likely to be potentially 'over-supplied', for a given handling capacity, by local farms.

CHAPTER 7

Promoting investment for increased agricultural production and productivity

What emerges from the earlier sections is that:

i. Agricultural development depends on the simultaneous growth of agricultural production and the value chains to which it is linked.

ii. It needs to be recognized that there are different types of investors operating in agriculture and its value chains, including small and family-owned farms, large commercial farmers, domestic corporate sector players, transnational corporations, sovereign wealth funds, the public sector (international, regional, national and local governments) and donors, and that they all have different objectives and roles.

iii. There is also a variety of sources of financing for investment, including savings and retained profits; credit from formal and informal sources; equity financing; donations and grants; FDI and government revenues. These sources are not equally accessible to all investors.

iv. On-farm capital owned by farmers is decisive for agricultural production and labour productivity. Other forms of agricultural investment are likely to have limited or no impact if they are not accompanied by increased on-farm investments. On-farm capital depends on farmers' own investment of labour and financial resources, which are generated primarily from their own savings.

v. Public investment, though relatively small, is the second most significant contributor to farm-level capital formation, both directly through the provision of rural public goods, and its complementarity effect on private investment.

Therefore, to increase farm-level investment it is imperative to:

i. Promote farm household savings.

ii. Increase public expenditure and investment in agriculture.

iii. Create an enabling environment for corporate private sector investment in agro-industries.

7.1 PROMOTING FARM-HOUSEHOLD SAVINGS FOR ON-FARM INVESTMENT

As noted earlier, the case studies showed that 'household savings' are the main source of investment by the farmers, and in the presence of credit constraints, the saving and investment decision of an agricultural household can hardly be separated since its investment decisions are linked to its saving decisions. As regards agro-industries, it was also noted that globally firms finance about two-thirds of their investments from retained earnings and another 16–23%, depending on the size of the firm, from bank loans (UNCTAD, 2008).

The High Level Committee on Estimation of Saving and Investment in India (2009), after estimating the investment gap in Indian agriculture stated that: 'Given the imperative need for enhancing the level of investment in the farm sector, concerted efforts have to be made for generation and mobilisation of savings in order to realise the desired growth potential of the agriculture sector'.

The central idea of Lewis's (1955) traditional development theory was that increasing savings would accelerate growth. Conversely, several more studies have concluded that economic growth contributes to savings (Sinha and Sinha, 1998; Salz, 1999; Caroll *et al.*, 2000; Anoruo and Ahmad, 2001).

Although debate is ongoing on the role of savings and investment in promoting economic growth, economic history suggests that countries that were able to accumulate high levels of domestic investment, largely financed by domestic savings, achieved faster rates of economic growth and development. A study of 32 countries by Krieckhaus (2002) notes that a higher level of national savings led to higher investment and consequently, higher economic growth. *The Growth Report: Strategies for Sustained Growth and Inclusive Development* (Commission on Growth and Development, 2008), investigated 13 countries, classified as most successful in achieving growth and development, in order to identify the reasons for their success. The Report noted that, among other things,[18] they all mustered high rates of saving and investment, including public investment in infrastructure. They were all 'future-oriented', forgoing consumption in the present in pursuit of a higher level of income in the future. In the mid-1970s, Southeast Asia and Latin America had similar savings rates. Twenty years later, the Asian rate was about 20 percentage points higher. China has saved more than a third of its national income every year for the past 25 years. This saving has been accompanied by prodigious rates of domestic investment.

Any economic activities that do not contribute to positive savings cannot grow and, as some savings are essential to cover depreciation to maintain the existing capital stock, are by definition not sustainable. The process of economic growth hinges critically on the generation of greater savings and directing it toward productive investments.[19]

One of the rationales for development cooperation, particularly ODA, was guided by the savings gap of the developing countries. It is argued that developing countries need more capital. However, since the creation of capital is a function of the level of savings, developing countries face the dilemma of not having enough capital precisely because they are too poor to save. In other words, their savings are insufficient to free up a part of their domestic potential for the production of capital goods or for the production of exports that could finance imports of such goods. In this theoretical framework, until economies reach critical benchmarks of savings and investment, they are not expected to grow fast enough to initiate a catching-up process (see, for example, Rosenstein-Rodan, 1961 and Sachs *et al.*, 2004). The attempt to fill this 'savings gap' by capital inflows from countries with higher income and savings guided traditional development thinking. The idea was that the developed country

---

[18] According to the Report, a close look at the 13 cases reveals five striking points of resemblance. These are: (i) they fully exploited the world economy; (ii) they maintained macroeconomic stability; (iii) they mustered high rates of saving and investment; (iv) they let markets allocate resources; (iv) they had committed, credible, and capable governments.

[19] It should, however, be noted that savings affect growth positively in those countries that are not too close to the technological frontier, but does not affect it at all in countries that are close to the frontier. Growth may result from innovations that allow the domestic sectors to catch up with the frontier technology (Aghion *et al.*, 2006).

should help developing countries until their marginal saving rate exceeds the average. At that stage foreign assistance can end, as developing countries will be able to break away from the vicious cycle of poverty and enter the virtuous cycle of save, invest and grow.

It should be noted that there are no unique set of policies for increasing savings and domestic investment in agriculture. Savings and investment can be both the trigger for and the outcome of specific national policies, strategies and institutional arrangements. A wide range of policy issues and institutional factors (legal institutions, socio-cultural values, governance, land tenure, property rights, natural resource management, education, R&D, infrastructure) affect savings and investment.

However, there are several essential requirements that must work in tandem with each other and with sectoral and overall policies. Fulfilling only one of these requirements without considering the others is not likely to be effective in promoting investment.

The essential requirements

Promotion of savings by ensuring their ownership, transferability and transformability through good governance and rule of law

People save to transfer and eventually transform their savings into capital (Alamgir, 1976). For this process to function efficiently, good governance and the rule of law are required. For savers, it is important to ensure that whatever is saved and transformed into capital is legally owned and that there is a system based on the rule of law that can guarantee ownership and transferability in case of dispute. To be effective, the legal system must be equally accessible and affordable to all.

Property rights, fixed capital and financial institutions

All the different agents of the economy (public sector administrations, private corporations, small businesses, including farms, and individual households) have different savings and investment behaviour. However, for households, one of the primary motivations for saving is the formation of fixed assets and capital. With fixed capital, farmers gain access to financial markets and can borrow working capital for further investments. No financial institution lends without collateral. For this reason, fixed capital formation is a driving force for economic growth, development and the reduction of poverty and hunger. The crucial factors that allow for the formation of fixed capital are clearly defined property rights that are applied fairly and equitably to all under the rule of law and the presence of working financial institutions.

To transform savings into fixed assets and capital along with property rights, there is a need for financial institutions. Without financial institutions, which enable the transformation of savings, savings will not enter the economy for investment and form part of the virtuous cycle of 'save, invest and grow'. In addition to being a determining factor for investment in fixed capital, property rights are also a determining factor in the transformation of fixed capital into financial capital. De Soto (2000) states that capital is the force that raises the productivity of labour and creates the wealth of nations. It is the lifeblood of the capitalist system and the foundation of progress. Capital is the one thing that the poor countries of the world cannot seem to produce for themselves, no matter how eagerly their people engage in all the other activities that characterize a capitalist economy. De Soto demonstrates, with the help of facts and figures collected block by block and farm by farm in Asia, Africa, the Middle East and Latin America, that even in the poorest countries, the poor save, but they hold these resources in forms that inhibit investment: houses built on land whose ownership rights are not adequately recorded and unincorporated businesses with undefined liability. Because the rights to these possessions are not adequately documented, these assets cannot readily be turned into capital, cannot be traded outside of narrow local circles where people know and trust each other, cannot be used as collateral for a loan, and cannot be used as a share against an investment.

According to the World Bank (2004), secure rights to land encourage investment. Farmers in Thailand with title to their land invested so much more in their land that their output was 14–25% higher than those working untitled land of the same quality. In Viet Nam, if rural households have a document assigning clear rights of control and disposition of resources, they commit 7.5% more land than households without documentation to crops that require a greater initial outlay and bring in yield after several years. In Peru, almost half of those with title to their property in Lima's squatter settlements have invested in improvements, compared with 13% of those without title.

The same World Bank report also states that titling can improve access to credit when markets, lending institutions and the other elements of a financial infrastructure are present. Land ownership is an important indicator of creditworthiness, and a registered title allows lenders to easily verify ownership. Titled land is also accepted more readily as collateral. Lenders can determine whether others have an interest in the property and assess the likelihood of seizing the land if the borrower refuses to repay the debt. In Costa Rica, Ecuador, Honduras, Jamaica, Paraguay and Thailand, farmers with secure title obtain larger loans on better terms than those without secure title. In Thailand, farmers with title borrowed anywhere from 50% to five times more from banks and other institutional lenders than farmers with land identical in quality but without title. The benefits of secure land entitlement extend beyond the farm. In Peru, residents in Lima that received title to their land have used the titled land as collateral to buy microbuses, build small factories and start other types of small businesses. Lack of secure title, which is common in many countries, is thus one more obstacle the smaller entrepreneurs face when trying to finance their operations. Securing rights contributes to a better investment climate in more ways than simply boosting investment and easing access to credit.

Securing rights benefits both individual landholders and the community at large by creating investment incentives and improving credit access, which foster economic growth and reduce poverty. The fees governments charge for titling services should reflect this mix of public and private benefits. They should not be inflated through excessive red tape or demands for 'unofficial' payments by registry staff. Surveyors, notaries, and other professionals who have a monopoly on the preparing the necessary documentation can also boost costs and deter registration. In Russia, surveyor fees equal to 2 years of the minimum wage and keep many from registering their property. In Peru, breaking the notaries' monopoly over drafting deeds was crucial to the titling of urban land belonging to the poor and near-poor.

The World Bank also recommends considering alternatives to full-blown titling. A large titling programme is costly and requires many trained professionals. Before initiating a programme, governments should consider whether their policy objectives can be realized through measures that do not provide full legal title. Indeed, experience around the world shows that a diversity of tenure options can facilitate access to land. In Niger, security of rights was realized through a simple, community-based registration scheme. In Honduras, simple title documents that lenders can hold while the loan is outstanding have been enough to improve the flow of formal credit to small farmers. In urban areas interim measures short of full titling can also begin to meet residents' needs for greater security. Botswana has issued land use certificates that have protected holders from eviction while the government considers options for addressing urban landlessness. India and Viet Nam provide other examples where secure rights to land were achieved without requiring a full-blown title to the property.

Promotion of savings and income by allowing and facilitating land consolidation for a level of income that is adequate for saving

In most countries with acute food insecurity and poverty, most of the smallholder farmers are not in a position to save. As has already been noted, for farmers, the main sources of investment finance are their own savings and their fixed capital, which is used as collateral for credit. Capital formation is certainly higher for farming households with positive savings and clear, legally recognized ownership of their land. The same is true for farmers with larger than average land holdings, more fixed assets and more diversified production. However, in countries where the levels of poverty and hunger are high, such as India and Bangladesh, the average farmer does not earn half of what is needed to cross the poverty line. For small and marginal farmers with below average land holdings, the situation is even worse, both in terms of their ability to save and secure their rights to the land.

In Bangladesh, according to the Household Income and Expenditure Survey, 2010 (Bangladesh Bureau of Statistics, 2010), households with no land and households with land but with holding size of less than half a acre (nearly 66% of the holdings of the country) have negative savings. In addition, households with land holdings size up to 1.5 acres do not have any positive savings. It would appear then that all households with less than 1.5 acre of land, nearly 85% of farming households, including landless farmers, do not have any positive savings.[20]

A remarkable feature of the Indian macro-economic story since independence has been the continuous rise in household savings. Since the 1950s, the household sector has remained the predominant source of gross domestic savings (GDS). It contributed, on an average, around 74% to total domestic savings between 2002 and 2007. However, at the same time the estimated annual savings of all farming households during 2002–03 was found negative at 69,348 Indian rupees. As a result, the ratio of farm sector savings to overall GDP is estimated at –2.8% for the year 2002–03 (High Level Committee on Estimation of Savings and Investment, 2009).

To promote farm-level investment, land consolidation needs to be facilitated to enable farmers to attain a level of income adequate for positive savings. In this context, it bears repeating that economic activities that do not contribute to positive savings cannot grow and by definition cannot even be sustainable, as some savings are essential to cover depreciation and maintain the existing capital stock. Land consolidation, however, needs to be supported by the generation of non-farm income opportunities.

---

[20] Based on: savings = income − expenditure. However, the situation may be different once some types of expenditure are taken into account, such as those on housings and the purchase of durables as investment.

7.2 PROMOTING PUBLIC SECTOR INVESTMENT IN AGRICULTURE

The level and composition of public investment is considered to be determined by political economic concerns. For this reason, the need to improve budget processes is receiving increased attention (World Bank, 2011). However, there is also a need to look at the policy-making process because policies affect, in an integrated way, both private and public investment in and for agriculture. Appropriate policies can enhance the returns to both private and public investment. Appropriate public investment can also enhance returns to private investment and improve incentives to invest. An inappropriate policy framework can significantly reduce the impact of investments and lead to substantial waste of public resources.

The political economic considerations that influence agricultural policy choices include: (i) interest group and collective actions; (ii) interaction of voters and interest group with politicians; (iii) the type of political regime; (iv) social mobilization; and (v) ideas and ideology (Birner and Resnick, 2010). Considerable emphasis in the literature has been given to 'interest group approaches', which focus on differences in the ability of rural and urban groups to overcome the difficulties of engaging in the collective action and organizing themselves into interest groups that can exert political pressure (Olson, 1965; Birner and Resnick, 2010).

According to Birner and Resnick (2010), ideas and ideology are typically considered endogenous in the economics literature on agricultural policy choices, based on the assumption that ideas and ideologies are merely used to defend economic or political interests. As a result, few quantitative political economy models account for ideology (de Gorter and Swinnen, 2002). However, the qualitative literature suggests that ideas and ideologies play an important role in explaining agricultural policy choices. In the pre-structural adjustment period, policies that taxed agriculture to finance rapid industrialization have been influenced by the notion of 'African socialism' and by dependency theory (Krueger *et al.*, 1991). These policies were also supported by the mainstream economic thinking of the time, as emphasized by Bates (1981). The withdrawal of public sector intervention in agriculture in the structural adjustment period was obviously driven by a general paradigm shift in international development thinking (Paarlberg and Grindle, 1991). Likewise, the current discussion on agricultural policies that support smallholder farmers, such as 'market-smart subsidies', is influenced by the 'Post-Washington Consensus'.

A country's desire to be self-sufficient, either based on its own production or on trade plays a dominant role in agricultural policy discourse and the allocation of public investment in agriculture. Self-sufficiency has been the driving force affecting public expenditure and policies affecting trade in agricultural products in many developing and developed countries. The political orientation of a regime along the political spectrum from left to right, and from nationalist to globalist, can also be assumed to influence agricultural policy choices, the allocation of public resources and the priorities of the overall economic and sectoral development strategies.

Many countries are currently making concrete efforts to guide and improve investment in agriculture by developing country investment plans (CIPs). However, these CIPs are largely determined by the country's overall economic and sectoral development strategies. As noted earlier, these strategies are influenced by government priorities that are shaped by its ideology and political economic considerations.

It is important that the development strategy leads to the adoption of policies and programmes that will contribute to increasing farm household savings and investment and contribute to strengthening the virtuous cycle of save, invest and grow.

An appropriate agricultural development strategy for promoting investment

As noted earlier, trends in the reduction of policy biases against agriculture, increasing globalization and urbanization, and higher food and energy prices are making agriculture a more profitable business opportunity for private sector engagement. There has been a surge in small-, medium- and corporate-scale private sector investment in agricultural value chains. Some governments and donors are actively encouraging these value chain investments through various kinds of agribusiness support programmes.[21] Private sector investments along value chains are opening up new market opportunities for some farmers. However, it is also becoming apparent that many small farms are being left behind. While some are being successfully pulled up by new value chain opportunities, many others are sinking into deeper poverty and subsistence modes of production because of higher food prices and reduced access to land and markets.

This polarization between 'commercializable' and 'non-commercializable' subsistence agriculture is more muted in countries where small farms can invest and/or diversify their livelihoods out of farming (e.g. in some fast growing Asian economies). However, there are many instances where this is not yet possible on the scale required or where the returns to non-farm activities remain too

---

[21] Some of these programmes are overtly aimed at promoting a more 'industrialized' agricultural sector even where, as in India, this must de facto be based largely on small farm production.

low (e.g. much of Africa). Unless carefully managed, this polarization could lead to a situation where policies and investments geared towards strengthening commercial agricultural production and value chains are not consistent with policies and investments geared to reducing poverty and food insecurity.[22]

Farmers with viable market prospects and capable of generating positive savings need to be supported as a business proposition. Farmers that face a stagnant or worsening situation, without any opportunity or possibility to save, need to be supported in more humanitarian ways that help them become more food secure and allow them to successfully diversify out of farming. Some kinds of interventions are beneficial to both groups of small farms (e.g. rural roads and some kinds of R&D). Many others need to be tailored differently to the first group of farmers (e.g. credit and insurance, marketing support and cash transfers). In the past, the need for more sharply differentiated small-farm strategies has not been so necessary. The need for a differentiated strategy has important implications for how policies and investments to improve food security, reduce poverty, develop agribusiness opportunities and promote agricultural growth in general should be structured and integrated.[23] As noted earlier, farmers who are unable to save are also unable to invest and public sector support and investment are not a substitute for the investment that farmers themselves need to make to increase production. Providing support to farmers without savings to gain access to credit often contributes to their indebtedness. It can even increase the number of poor and hungry.

These considerations suggest the need for a three-prong strategy for pro-poor agricultural growth (FAO, 2012d):

i. Promote the growth of commercial agriculture and its value chains, using public policy to enable the private sector (farmers and agro-industrialists) to take the lead (described in the next section).

ii. Shape that engagement in ways that enable as many small farms as possible to link to markets and successfully commercialize by investing their own savings.

iii. Put in place support programmes targeted to those small farms that cannot

---

[22] The strategy of focusing public sector investment in 'growth poles', 'clusters', 'corridors' and the like in order to encourage the private agro-industry sector to invest in businesses linked to (well-supported, serviced and organized) small farm production will exacerbate this polarization effect.
Commercial agriculture and value chain development may well enhance urban food security while doing nothing for rural food security, possibly even diminishing it in some circumstances.

[23] Among the reasons for this emergence are: (i) the substantial increase in global commercial capital seeking exposure to agriculture and agribusiness given the recent widespread perception that it is now likely to be significantly more profitable than it used to be for the same degree of risk; and (ii) technological advances that have improved the economies of scale for large farms and also made it possible to introduce sustainable cropping into agro-ecological zones previously considered too fragile or marginal to support intensive agriculture.

succeed as viable businesses. Key policy recommendations that relate to each part of this strategy are summarized below.

Promote the growth of commercial agriculture and its value chains

- Create an enabling business environment for private investment at all stages of the value chains. This includes: macro-economic and trade policy; liberalization of domestic markets; taxes; regulations; price policies; subsidies; and contract enforcement. There is need for stability in policies over time. Avoid agricultural policies that crowd out the private sector, such as state control of fertilizer procurement and distribution.

- Prioritize those rural public investments that complement rather than substitute for private investment: rural infrastructure, especially roads; some types of agricultural and agro-industrial R&D; electrification; education; and health.

- Promote on-farm investment by reforming property rights, supporting NGOs and others groups that facilitate collective action in rural communities, and strengthening rural financial services.

- Support programmes for agro-industry (e.g. industrial clusters, public–private partnerships, innovative financing, improved access to land and water).

Facilitate the commercialization of more small farms

- Ensure that the enabling policies and public investments mentioned above reach large numbers of small farms. This may involve going beyond breadbasket areas to invest more in less developed regions and communities, where the returns can be high for both growth and poverty reduction. Problems with property rights and the need for collective action are often greatest in these areas.

- Ensure that rural financial services, including insurance, serve small farms. This may require subsidies, or public–private risk sharing through mechanisms such as credit guarantees and warehouse receipt systems.

- Promote contract farming arrangements and the growth of producer organizations that can serve as market intermediaries for small farms.

- Establish targeted training programmes, especially for women farmers and young farmers.

- Tie support programmes for agro-industry to commitments to engage with small farms and SMEs, and complement these programmes with targeted subsidies and tax breaks.

Support programmes for non-viable farms

- Because it is not easy to identify farms that are not going to be viable as businesses, priority should be given to support measures that encourage self-selection or involve community leaders in targeting the selection (Grosh *et al.*, 2008).

- Provide training and support for income and employment diversification out of farming, such as non-farm business or employment, or migration.

- Provide training, technical support and perhaps targeted subsidies to encourage food gardens and support low-cost intensification of on-farm productivity of food crops, especially for women farmers.

- Establish partnerships with NGOs, some of which can be particularly effective for delivering these interventions.

- Establish and sustain well-designed safety nets, such as employment programmes and emergency relief (Grosh *et al.*, 2008).

7.3 CREATE ENABLING ENVIRONMENT FOR CORPORATE PRIVATE SECTOR INVESTMENT IN AGRO-INDUSTRIES

For all businesses, the primary motivation for investment is to increase the capacity of the enterprise to survive and prosper. Businesses do this by exploiting opportunities to grow total profits either by increasing sales or profitability or both. To the prospective investor, some business environments will appear inherently more attractive than others, regardless of the specific investment opportunity under consideration. Key factors shaping this perception are the nature and magnitude of the risks to which the investment will be exposed and the amount of capital that will be required. The amount of capital is largely a function of the minimum necessary scale required for the core agro-industrial activity and the amount of investment in ancillary infrastructure and services required to support this activity.

In the context of agro-industrial development, Christy *et al.* (2009) identify a hierarchy of enabling requirements that governments can consider in addressing its role in enhancing the business environment. Using a scale that represents a cumulative progression from necessary to sufficient conditions, the authors describe three types of business enablers: 'essential', 'important' and 'useful'.

Essential enablers

Essential enablers make possible the functioning of markets and enterprises. They include land tenure and property rights, infrastructure, and domestic and foreign trade policy. In this respect, the 'wrong' business environment has in the past been a significant deterrent to FDI. For example, companies have made high-level decisions against investing in certain countries because of the risks and costs involved. A high degree of uncertainty about government intentions and likely actions is always a major deterrent to investment. In a number of regions, there is a correlation between economic reform, particularly market liberalization, and agri-food investment and value-chain financing. More specifically, the ease with which access to land and water resources can be acquired and maintained is a major factor in many agro-industrial investment decisions. In practice, this access depends on widespread recognition of the corporate property rights, both formal (the application of legislation) and informal (acceptance of the local community). Similarly, the difficulties that SMEs face in ensuring contract compliance by their suppliers and buyers in situations of weak business governance is a significant barrier to their accessing credit and finance from the formal banking sector.

Good infrastructure is an essential underpinning for successful agro-industrial development. In countries with severe public budget constraints, it may pay to concentrate infrastructure (and services) within a particular area to enhance connectivity in agricultural value chains. Examples of such 'agroclustering' include food and agroprocessing parks and agricultural growth corridors and, more generally, special economic and free trade zones.

Important enablers

Important enablers are characterized by second-order activities that the state can and often does provide to facilitate commerce. These include norms, standards, regulations and services relating to production, R&D, and financial services for agro-industries. In general, the progressive development of laws, regulations and institutions that ensure agricultural marketing is conducted according to 'open market' principles. Of particular relevance in this regard are transactional transparency, competitiveness and reference to standardized product and delivery specifications, all of which are fundamentally important to agro-industrial development. As such, adhering to these principles can be considered both an essential and an important enabler, as agro-industrial value-chain development is often critically dependent on the ability to demonstrate quality control and origin traceability for the products. The export of fresh horticulture products is the classic example of this, but it is increasingly significant to relatively highly processed agricultural products, such as refined palm oil and chocolate.

The relevance of R&D to agro-industry lies primarily in its contribution to raising agricultural productivity and adaptability. This reduces the domestic resource costs of producing the raw material needed by agro-industry and enhances the potential market competitiveness of the final product. As da Silva *et al.* (2009) note, there is a disconnect between this product development in agriculture, most of which has been publicly funded through the international network of research institutes overseen by CGIAR and by national agricultural research organizations, and its commercialization, which is undertaken mainly by the private sector.

Agribusiness and biotechnology companies have been making very substantial investments in R&D for seed improvement, but this has mainly been for the large-scale, corporate farming market rather than for smallholder farming. Technological advances in agricultural value chains downstream from farm production, in processing, transporting, storing and packaging, have come from a combination of publicly and privately financed R&D, with the latter predominating in more recent years.

It is difficult to overstate the importance of effective financial services for agro-industrial development. One reason for the growing dominance of large-scale agribusinesses is their ability to organize and mobilize finance for large segments of agricultural value chains, often for entire chains. Much of this extends to smallholder farmers in the form of credit (in cash or kind) for input purchases provided by input suppliers and output purchasers. If SMEs are to be encouraged to invest more in agricultural value chain development, they will need more effective ways of hedging against risk to attract finance themselves. Futures markets require contract enforceability and dependable information systems. Based on a review of 20 case studies of agribusiness investment promotion, FAO identified the creation of specific financial mechanisms to facilitate agribusiness development (e.g. credit guarantees, agricultural/rural investment funds, and credit and savings products) as one of four promising approaches. The FAO review also made reference to an initiative in Republic of Korea where agricultural credit was funded half by an agricultural bank and agro-industry (managed by a private company) and half by the government.

Useful enablers

Useful enablers include the ease of doing business in a country, the business development services available to prospective investors, and the general intensity and effectiveness of business linkages (both horizontal and vertical) between enterprises in value chains. These are all areas where government policies and programmes, including specific investment promotion initiatives, can make a considerable difference to the attractiveness of the agro-industrial investment climate. Investment promotion agencies (IPAs) typically provide prospective entrepreneurs and investors with information and advice on markets and market opportunities. They also effect introductions into value chain networks, arrange access to finance and provide guidance on regulatory and compliance matters. IPAs may also provide finance directly for feasibility studies and even start-up or seed capital. Business 'incubators' are effectively IPAs that offer a comprehensive package of such services in particularly close relationship with entrepreneurs and require a detailed description of their strategies or business model. An important objective is to reduce the level of risk perceived by entrepreneurs and investors.

Business development services are potentially useful to agro-industrial SMEs. These often operate in an environment of substantial uncertainty with limited knowledge of how to proceed. Consequently, they lack access to technology, finance and market information. However, by themselves business development services are unlikely to have much impact and they are best provided as part of a package that provides access to these key resources.

7.4 PROMOTING FOREIGN DIRECT INVESTMENT[24] THROUGH INCLUSIVE BUSINESS MODELS

To take advantage of the opportunities offered by FDI for the promotion of agro-industries, while minimizing risks, developing country governments should promote the use of win-win business models for investment. The case studies suggest that investment projects that give local farmers an active role and leave them in control of their land tend to have positive effects on local economic and social development. Successful projects combine the strengths of the investor (capital, technology and expertise in management and marketing) with those of local farmers (labour, land, traditional know-how and knowledge of the local conditions). This combination can provide the basis for win–win outcomes. Business models that leave farmers in control of their land give the farmers an incentive to invest in improving their land. Since the bulk of agricultural investments come from farmers themselves, these models are more likely to raise the level of agricultural investment in developing countries. However, it is important for governments to keep in mind that FDI only accounts for a small share of total agricultural investment. National policies should give more emphasis to increasing domestic investment, in particular by farmers, as they account for the bulk of investment in agriculture.

---

[24] This section draws from FAO, 2013b.

Large-scale land acquisition

Over the last 4 years many analysts, development agencies, NGOs and the media have focused on one specific category of primary agricultural investment: the acquisition of agricultural land on a large scale by foreign companies and sovereign funds. The focus on this type of investment is due to the numerous economic, political, social and environmental implications of land acquisition, especially if it is done by foreigners or for them. Due to methodological differences, the estimates provided by various sources vary substantially as to the area acquired by foreign firms. The more reliable cross-checked figures are not as high as many reports in the media suggest. Nevertheless, they do show that foreign investment in agricultural land in developing countries has increased markedly over the past decade. More importantly, the lands acquired by foreign investors tend to be among the best ones, with good soil quality, high production potential, irrigation and proximity to infrastructure and markets. As a majority of foreign investment projects aim at export markets or the production of biofuels, they may pose a threat to food security in low-income food-deficit countries. This is especially true if they replace food crops destined for local markets. The net effect on food security will also depend on the additional income generated by the project, its sustainability and how this income is distributed in the local economy.

Large-scale acquisition of agricultural land can have other adverse impacts, especially in countries where there is a lack of good governance, rule of law, transparency and clear land tenure rights. These negative effects include the displacement of smallholder farmers, the loss of grazing land for pastoralists, the loss of incomes and livelihoods for rural people, and the depletion of productive resources. In addition, large-scale land acquisition can have a general negative impact on livelihoods due to reduced access of local people to resources. This can lead to social fragmentation. There is also evidence of adverse environmental impacts, in particular the degradation of natural resources such as land, water, forests and biodiversity. The case studies show that when such impacts arise they generate opposition to the project by local people. Sometimes, this leads to the occupation of part of the land or hostile actions such as vandalism. Opposition can force the investor to engage in costly and time-consuming litigation and lawsuits. It also increases transaction costs and reduces the return to the investment. The negative effects are likely to be worse when the company only utilizes a small share of the land it has acquired in areas where land is high in demand.

While a number of studies document the negative impacts of large-scale land acquisition in developing countries, there is much less evidence of its benefits to the host country, especially in the short term and at the local level. The main type of benefits appears to be the generation of employment, but there are questions as to the sustainability of the jobs that are created. In several projects, the number of jobs has decreased over time and was lower than the investor's initial announcements. There is also the issue of the quality of the employment created and who benefits. Managerial positions tend to be occupied by expatriates or persons originating from areas other than where the project is located. In some projects, even low-skilled worker jobs were mainly taken up by non-locals. Another expected advantage of FDI in developing countries is the transfer of technology. In the case of large-scale land acquisition the evidence for this is mixed. Obviously assessing the extent of technology transfer and other outcomes of the investment takes time. It may be that the investment was too recent for the transfer of technology to have occurred or to be observed.

In conclusion, the studies suggest that for investments involving large-scale land acquisitions in countries where land rights are unclear and insecure, for local communities, the disadvantages often outweigh the benefits, especially in the short run. This outcome is even more likely when the acquired land was previously utilized by local people in either a formal or informal manner. Consequently, acquisition of already utilized land to establish new large farms should be avoided and other forms of investment, with a focus on agro-industries, should be considered. Even from the investor's perspective, business models that do not involve the transfer of the control of land are likely to be more profitable.

There is a broad variety of inclusive business models for agricultural development. Studies suggest that none of them can be presented as the ideal approach to agricultural development in all contexts. There is no one size fits all business model. Different situations will require different models. Local economic and social factors, including the level of organization of the community, the strength of local institutions, the technical level of farmers and the effectiveness of their organizations will determine the type of model that is most likely to succeed. In cases where farmers are unable or reluctant to create an organization, contract farming may be the most appropriate model. Conversely, in communities where there is a strong tradition of collaboration and effective farmer organizations, an out-grower scheme giving farmers a share of the capital, or possibly a joint-venture between the investing company and a farmer cooperative, may be the most appropriate option. Other factors that condition the success of business models include the national legal and institutional framework; the specific terms and conditions of the investment contract; and the experience, skills and motivations of the investor.

International guidance

Among the many factors that determine the impacts of foreign investment on the local economy, the domestic laws and institutions governing agricultural investment and land tenure are critical. However, in developing countries they are often inadequate to ensure sustainable agricultural development, especially in terms of legal enforcement. Developing country governments and local institutions need support in the form of policy advice, capacity building and technical assistance. Useful guidance can be obtained from some recent international agreements. Of particular importance are the *Voluntary Guidelines on the Responsible Governance of Tenure of Land, Fisheries and Forests in the Context of National Food Security (VGGT).*[25] The Committee on World Food Security (CFS) adopted the Guidelines in May 2012 after 3 years of international consultations involving governments, civil-society organizations and companies. The VGGT serve as a reference and provide guidance to improve the governance of tenure of land, fisheries and forests. Their overarching goal is to ensure food security for all. One of the implementation guides currently under development deals with agricultural investment. Another important internationally agreed instrument is the *FAO Voluntary Guidelines on the Right to Food.*[26]

The CFS has also recently launched a consultation process for the development of principles for responsible agricultural investment that enhance food security and nutrition. It is expected that the principles resulting from the consultations will have international recognition and serve to guide agricultural investment. They will refer to and build on the VGGT. The consultations will take into account various existing instruments, including the voluntary principles for responsible agricultural investment that respect rights, livelihoods and resources (PRAI).[27] The PRAI have been jointly formulated by FAO, IFAD, UNCTAD and the World Bank to serve as a possible reference framework for governments in the development of national policies, laws and regulations, or in the negotiation of international investment agreements and individual investment contracts. The PRAI are a set of very general principles that need to be translated into more operational guidance. To this end, the World Bank and UNCTAD are implementing pilot projects with governments, investors and civil-society organizations in selected developing countries. The results of the projects will feed into the CFS consultations.

---

[25] www.fao.org/nr/tenure/voluntary-guidelines/en/
[26] www.fao.org/righttofood/publi_01_en.htm
[27] www.responsibleagroinvestment.org

ANNEX 1.
Sources of investment finance, selected country groups, 2002–2006

	Number of countries	Number of firms	Internal funds and retained earnings	Local and foreign-owned commercial banks	Investment and state funds (a)	Trade credit	Equity	Family and friends	Other
					%				
All countries									
All	100	32,809	65.5	16.1	1.3	3.2	3	3.8	7.1
Small	100	12,388	69	12.4	1.1	3	3.4	4.7	6.4
Medium	100	11,235	63.1	17.9	1.5	3.4	3.4	3.1	7.7
Large	100	9,036	59.7	22.9	2.5	3.4	2.9	1.5	7.1
Developed countries									
All	5	2,592	59.3	20	0.6	3	3.8	1.2	12
Small	5	1,618	63.2	18.1	0.3	2.7	3.2	1.7	10.9
Medium	5	575	53.4	22.8	0.8	3	5	0.4	14.5
Large	5	399	50	25.5	1.5	3.4	5	0.5	14.2
Emerging-market economies in Europe									
All	8	2,334	59.6	13.9	1.1	2.4	7.4	2.5	13.1
Small	8	1,290	62.8	10.1	0.2	2.8	7.5	4.2	12.3
Medium	8	621	55.3	18.3	1.4	2.4	8.2	0.4	14
Large	8	423	57.8	18	3	1.4	6.5	0.1	13.2
Latin America and the Caribbean									
All	20	7,845	60.6	20.2	1.5	6.8	1.2	2.7	7
Small	20	2,622	62.2	18.6	1.1	6.4	0.8	3.2	7.8
Medium	20	3,265	58.9	21.2	1.1	7.6	1.6	2.8	6.9
Large	20	1,938	58.8	24.4	2.8	6.3	1.1	1.3	5.3
Africa									
All	31	6,100	73.8	12.7	1.3	2.1	0.8	3.7	5.6
Small	31	2,642	77.8	8.9	1.1	2.4	0.8	4.3	4.8
Medium	31	2,059	69.9	16.1	2	1.9	1	2.5	6.6
Large	31	1,372	63.4	24.3	2	2.3	1.1	0.8	6.1
East, West, South and Southeast Asia									
All	17	9,309	49.3	21	1.6	2.8	8.9	7.2	9.3
Small	17	2,055	53.4	14.4	2.1	2.5	11.4	8.3	7.8
Medium	17	3,223	50.2	19.2	1.4	2.8	9.3	7.4	9.7
Large	17	3,928	46.4	25.9	2.8	3.1	8	5	8.8
Transition economies in Europe									
All	12	3,008	72.5	14.5	1	2.3	1.9	3.2	4.6
Small	12	1,448	77	10.4	0.4	1.7	2	5	3.5
Medium	12	915	69.8	16.5	1	2.5	2.3	2.5	5.4
Large	12	645	65.7	20.6	2.3	4.1	1.2	0.3	5.8
Transition economies in Central Asia									
All	7	1,621	81.4	10.1	1.9	1.3	0.2	2.9	2.2
Small	7	713	84.6	7.7	1	0.4	0	4.5	1.8
Medium	7	577	79.6	11.1	2	2.3	0.4	2.5	2
Large	7	331	77.8	14	3.1	1.2	0.1	1	2.8
Memo items: firm-based averages									
All		32,809	58.9	19.5	1.3	3.7	4.7	3.6	8.2
Small		12,388	67.7	12.5	0.7	3.5	4.2	4.9	6.4
Medium		11,235	56.8	20.6	1.4	4.3	4.8	3.4	8.7
Large		9,036	49.6	27.5	2.1	3.3	5.4	2.1	10
New		1,070	63.9	13.8	1.7	2.7	6	6.1	5.8

SOURCE: UNCTAD secretariat calculations, based on World Bank, *Enterprise Survey* database.

Note: New firms = firms aged 2 years or less. Small firms = less than 20 employees; medium firms = 20–99 employees; large firms = more than 99 employees. The numbers for small, medium and large firms may not add up to the total number given for all firms because some firms gave no indication of their size. Emerging-market economies in Europe: Czech Republic, Estonia, Hungary, Latvia, Lithuania, Poland, Slovakia and Slovenia.
(a): Aggregate funding by investment funds, development banks and other State services.

The NEPAD-OECD Draft Policy Framework for Investment in Agriculture

In recognition of the importance of an enabling environment for agricultural investment, the OECD and the New Partnership for Africa's Development (NEPAD) have developed a draft policy framework for promoting investment in the sector. The essential elements of this framework are given below. These include many of the issues raised in this paper, including the need for good governance and property rights. The Draft Policy Framework for Investment in Agriculture was prepared within the framework of the NEPAD-OECD Africa Investment Initiative and presented at the 5th NEPAD-OECD Ministerial Conference 26–27 April 2011. A summary of the issues addressed in this framework is as follows:

Investment policy: transparency of laws and regulations, property rights to land and other assets, protection of intellectual property and contract enforcement.

Investment promotion and facilitation: institutions and measures for promoting investment in agriculture, technology transfer to local farmers and public–private sector dialogue.

Human resource and skills development: human resource development, training of local farmers and local research and development capacity.

Trade policy: customs and administrative procedures, assessment of impact of trade policies, export promotion and financing, regional trade agreements.

Environment: policies for natural resource management and cleaner technologies, integration of R&D and environmental policies, energy needs and mitigation of extreme weather.

Responsible business conduct: labour standards in agriculture, enforcement of human rights, environmental protection, labour relations and financial accountability.

Infrastructure development: coherent infrastructure, rural development and agricultural policies, transparent funding procedures, information and communications technology for farming, incentives to private investment in secondary roads, water resource management and storage facilities.

Financial sector development: regulatory framework for agricultural finance, banking sector competition, functioning capital markets, instruments for risk mitigation, access to credit by local farmers and small and medium enterprises, guarantee and insurance mechanisms to support smallholders accessing credit and business development services for local farmers.

Taxation: tax policies supporting agricultural investment, appropriate tax burden on agribusiness, transparent and efficient tax policy and administration, coordination of central and local tax administration and funding of local public goods.

SOURCES: OECD, 2011.

ANNEX 3.
Non-financial assets in the UN System of National Accounts

At the first level of classification, there are four categories of non-financial assets. The first three categories are produced assets – fixed assets (611), inventories (612) and valuables (613) – and the fourth consists of all non-produced assets (614). The numeric codes pertain to the IMF GFS classification system are kept to better see the structure.

61	Non-financial assets
611	**Fixed assets**
6111	Buildings and structures
61111	Dwellings
61112	Buildings other than dwellings
61113	Other structures
61114	Land improvements
6112	Machinery and equipment
61121	Transport equipment
61122	Other machinery and equipment
611221	Information, computer and telecommunication (ICT) equipment
611222	Machinery and equipment not elsewhere classified
6113	Other fixed assets
61131	Cultivated biological resources
611311	Animal resources yielding repeat products
611312	Tree, crop, and plant resources yielding repeat products
61132	Intellectual property products
611321	Research and development
611322	Mineral exploration and evaluation
611323	Computer software and databases
6113231	Computer software
6113232	Databases
611324	Entertainment, literary, and artistic originals
611325	Other intellectual and property products
61133	Weapons systems
612	**Inventories**
61221	Materials and supplies
61122	Work in progress
61123	Finished goods
61124	Goods for resale
61125	Military inventories
613	**Valuables**
614	**Non-produced assets**
6141	Land
6142	Mineral and energy resources
6143	Other naturally occurring assets
61431	Non-cultivated biological resources
61432	Water resources
61433	Other natural resources
614331	Radio spectra
614332	Natural resources not elsewhere classified
6144	Intangible non-produced assets
61441	Contracts, leases, and licences
614411	Marketable operating leases
614412	Permits to use natural resources
614413	Permits to undertake specific activities
614414	Entitlement to future goods and services on an exclusive basis
61442	Goodwill and marketing assets

ANNEX 4.
A list of case studies

Country	Authors	Title
Bangladesh	Islam, S.M.F.	Human capital and capital formation in agriculture in Bangladesh.
Bolivia	Gutiérrez, M., Lünstedt, C. and Toranzo, C.	Operation of the Agricultural Sector: Capital Formation on and for Agriculture in Bolivia.
Bolivia	Hameleers, A., Antezana, S. and Paz. B.	Agriculture human investment strategies towards strengthening the farmers innovation capacity: Bolivia case study.
Brazil	Nascimento, J.R.	Analysis of international investments in the agricultural sector of Brazil.
Brazil	Santana, C.A.M. and Nascimento, J.R.	Public Policies and Agricultural Investment in Brazil.
Burkina Faso and Mali	Tshibaka, T.B. and Klevor, K.A.	Analysis of the Structure, Magnitude and Trends of Capital Formation in African Agriculture: the cases of Burkina Faso and Mali.
Cambodia	Hang, S.C., Socheth, H. and Chandarany, O.	Foreign investment in agriculture in Cambodia.
China	Huang, J. and Ma, H.	Capital Formation and Agriculture Development in China.
Egypt	Abdou, D.K., Taha, A.R.I. and El Shahat, A.I.	Structure, Magnitude and Trends of Capital Formation in and for Agriculture in Egypt.
Ethiopia	Atlaw Alemu, A., Meskel, A.G. and Yifredew, S.	Structure, Magnitude and Trends of Capital Formation in and for Agriculture in Ethiopia.
India	Bisaliah, S. and Mahendra Dev, S.	Private capital formation in Indian agriculture: an analysis of farm level data.
India	Mahendra Dev, S.	Macro and farm level investment in India: trends, determinants and policies. New Delhi.
Indonesia	Brighten Institute	Study on supportive policies for maximizing agricultural investment in Indonesia.
Indonesia	Brighten Institute	Corporate private sector investment in agriculture in Indonesia.

Laos	Douangsavanh, L. and Kunsy, P.	Agriculture investment trends: the role of public and private sector in Lao PDR.
Malawi, South Africa and Zambia	Chipeta, M.E.	Agricultural investment: approaches and country experiences of Malawi, South Africa and Zambia.
Malawi	Kumwenda, I.	Comparative experiences in agricultural funding for investment into capital formation and for other interventions to enhance agricultural production and productivity.
Nepal	Pant, K.P., Chettri, P. and Bhattarai, J.M.	Investigating the Structure, Magnitude and Trends of Capital Formation in and for Agriculture.
Republic of Korea	Global Agriculture Policy Institute	Investigating Supportive Policies and Resource Allocations for Agriculture Investment.
South Africa	Mdlulwa, Z.	Comparative experiences in agricultural funding for investment into capital formation and for other interventions to enhance agricultural production and productivity.
Paraguay	Nascimento, J.R.	Analysis of international investments in the agricultural sector of Paraguay.
Tanzania	Lyimo, B.	Sub Saharan Africa's Unfolding Tragedy in Mega Land Deals for Agro-investments with lessons from Tanzania.
Thailand	Waleerat, S. and Nipawan, T.	Analysis of International Investments in the Agricultural Sector of Thailand.
Viet Nam	Nga, N.T.D.	Agriculture investment trends: the role of public and private sector in Viet Nam.
Zambia	Sikombe, D.	Comparative experiences in agricultural funding for investment into capital formation and for other interventions to enhance agricultural production and productivity.

References

Aghion, P., Comin, D. & Howitt, P. 2006. *When Does Domestic Savings Matter for Economic Growth?* Working Paper 12275. Cambridge, MA, USA, National Bureau of Economic Research.

AgriFin. 2010. *Program Strategy. Agriculture Finance Support Facility.* Washington, DC, World Bank.

Ahmed, R. & Hossain, M. 1990. *Development Impact of Rural Infrastructure in Bangladesh.* IFPRI Research Report 83. Washington, DC, IFPRI.

Alamgir, M. 1976. Rural savings and investment in developing countries: some conceptual and empirical issues. *Bangladesh Development Studies* 4(1), 1–48.

Alexandratos, N. & FAO 1995. *World Agriculture: Towards 2010, an FAO Study.* Chichester, UK, J. Wiley & Sons, and Rome.

Alston, J. 2010. *The Benefits from Agricultural Research and Development, Innovation, and Productivity Growth.* OECD Food, Agriculture and Fisheries Working Papers No. 31. Paris, OECD.

Alston, J., Marra, M., Pardey, P. & Wyatt, T. 2000. Research returns redux: a meta-analysis of the returns to agricultural R&D. *Australian Journal of Agricultural and Resource Economics* 44(2), 185–215.

Anderson, K., ed. 2009. *Distortions to Agricultural Incentives: A Global Perspective, 1955–2007.* London, Palgrave Macmillan and Washington, DC, World Bank.

Anderson, K. & Brückner, M. 2011. Price distortions and economic growth in Sub-Saharan Africa. CEPR (Center for Economic Policy Research) Discussion Papers 8530. London, UK, CEPR.

Anderson, K. & Valenzuela, E. 2008. *Estimates of Global Distortions to Agricultural Incentives, 1955 to 2007.* Washington, DC, World Bank.

Anderson, K., Valenzuela, E. & van der Mensbrugghe, D. 2009. *Welfare and Poverty Effects of Global Agricultural and Trade Policies using the Linkage Model.* Agricultural Distortions Working Paper 52785. Washington.

Anoruo E. & Ahmad, Y. 2001. Causal Relationship between Domestic Savings and Economic Growth: Evidence from Seven African Countries. *African Development Review* 13(2), 238–249.

Athukorala, P. & Sen, K. 2001. *The Determinants of Private Saving in India.* ASARC Working Papers 2001–13. The Australian National University, Australia South Asia Research Centre.

Athukorala, P. & Sen, K. 2003. *Saving, Investment and Growth in India.* Oxford, UK, Oxford University Press.

Bangladesh Bureau of Statistics. 2010. *Household Income and Expenditure Survey 2010.* Dhaka.

Barre, M. 2006. *Investment as a Means to Agricultural and Rural Development in Africa: A Study on the Investment Needs for Agriculture and Rural Development.* Prepared for the FAO. Rome.

Barrett, C., Bachke, M., Bellemare, M., Michelson, H., Narayanan, S. & Walker, T. 2012. Smallholder participation in contract farming: comparative evidence from five countries. *World Development* 40 (4), 715–730.

Bates, R.H. 1981. *Markets and States in Tropical Africa – The Political Basis of Agricultural Policies.* Berkeley, University of California Press.

Benin, S., Nkonya, E., Okecho, G., Randriamamonjy, J., Kato, E., Lubade, G. & Kyotalimye, M. 2011. Returns to spending on agricultural extension: the case of the National Agricultural Advisory Services (NAADS) programme of Uganda. *Agricultural Economics* 42(2), 249–267.

Benoit-Cattin, M. 2011. Presentation at the FAO workshop on Policies for Promoting Investment in agriculture. Rome.

Birner, R. & Resnick, D. 2010. The political economy of policies for smallholder agriculture. *World Development* 38(10), 1442–1452.

Bisaliah, S., Mahendra Dev, S. & Syed, S. 2013. *Investment in Indian Agriculture: Macro and Micro Evidences.* New Delhi, Academic Foundation.

Bourdieu, P. 1986. The forms of capital. In: J. Richardson, ed. *Handbook of Theory and Research for the Sociology of Education*, pp. 241–258. New York, Greenwood.

Brookfield, H. 2008. Family farms are still around: time to invert the old agrarian question. *Geography Compass* 2(1), 108–126.

Caroll, C.D., Overland, J. & Weil, D.N. 2000. Savings and growth with habit formation. *American Economic Review* 90(3), 351–355.

Chaia, A., Dalal, A., Goland, T., Gonzalez, M., Morduch, J. & Schiff, R. 2009. *Half the World is Unbanked*. Framing Note. New York, Financial Access Initiative and McKinsey & Company.

Chand, R. 2000. *Emerging Trends and Regional Variations in Agricultural Investment and Their Implications for Growth and Equity*. Policy paper. New Delhi, National Center for Agricultural Economics and Policy Research.

Chand, R. & Kumar, P. 2004. Determinants of capital formation and agriculture growth. *Economic and Political Weekly* 39(52), 5611–5616.

Christy, R., Mabaya, E., Wilson, N., Mutambatsere, E. & Mhlang, N. 2009. Enabling environments for competitive agro-industries. In: C.A. da Silva, D. Baker, A.W. Shepard, C. Jenane & S. Miranda-da-Cruz, eds. *Agro-industries for Development*, pp.136–185. Rome, FAO and UNIDO (United Nations Industrial Development Organization).

Collins, D., Jonathan, M., Stuart, R. & Orlanda, R. 2009. *Portfolios of the Poor*. Princeton, USA, Princeton University Press.

Commission on Growth and Development. 2008. *The Growth Report: Strategies for Sustained Growth and Inclusive Development*. Washington, DC, World Bank.

da Silva, C., Baker, D., Shepherd, A.W., Jenane, G. & Miranda-da-Cruz, S. 2009. *Agro-industries for Development*. Rome, FAO and UNIDO.

Dastagiri, M.B. 2010. The effect of government expenditure on promoting livestock GDP and reducing rural poverty in India. *Outlook on Agriculture* 39(2), 127–133.

de Gorter, H. & Swinnen, J. 2002. Political economy of agricultural policy. In: B. Gardner & G. Rausser, eds. *Handbook of Agricultural Economics*, Vol. 2, pp. 1893–1943. London, Elsevier Science.

de Haas, H. 2010. Migration and development: a theoretical perspective. *International Migration Review* 44(1), 227–264.

de Soto, H. 2000. *The Mystery of Capital: Why Capitalism Triumphs in the West and Fails Everywhere Else.* New York, NY, Basic Books.

Deininger, K. 2003. *Land Policies for Growth and Poverty Reduction.* World Bank Policy Research Report. Washington, DC, World Bank.

Dercon, S., Gilligan, D.O., Hoddinott, J. & Woldehanna, T. 2009. The impact of agricultural extension and roads on poverty and consumption growth in fifteen Ethiopian villages. *American Journal of Agricultural Economics* 91(4) 1007–1021.

Dhawan, B.D. 1996. Trends and determinants of capital formation in Indian agriculture. *Indian Journal of Agricultural Economics* 51(4), April.

Dhawan, B.D. 1998. *Studies in Agricultural Investments and Rural Savings.* New Delhi, Commonwealth publishers.

Diakosavvas, D. 1990. Government expenditure on agriculture and agricultural performance in development countries: an empirical evaluation. *Journal of Agricultural Economics* 41(3), 381–390.

Diao, X., Fan, S., Kanyarukiga, S. & Yu, B. 2010. *Agricultural Growth and Investment Options for Poverty Reduction in Rwanda.* IFPRI Research Monograph, Washington, DC, IFPRI.

Domar, E.D. 1946. Capital expansion, rate of growth, and employment. *Econometrica* 14(2), 137–147.

Easterly, W. & Rebelo, S. 1993. Fiscal policy and economic growth: an empirical investigation. *Journal of Monetary Economics* 32(2), 417–458.

Evenson, R.E. 2001. Economic impacts of agricultural research and extension. In: B. Gardner & G. Rausser, eds. *Handbook of Agricultural Economics*, Vol. 1A, Chapter 11. Amsterdam, Elsevier.

Fan, S. 2008. *Public Expenditure, Growth and Equity: Lessons from Developing Countries.* Baltimore, Johns Hopkins University Press.

Fan, S. & Rao, N. 2003. *Public Spending in Developing Countries: Trends, Determination, and Impact.* EPTD Discussion Paper No. 99. Washington, DC, IFPRI.

Fan, S. & Saurkar, A. 2006. *Public Spending in Developing Countries: Trends, Determination and Impact* (mimeo).

Fan, S. & Zhang, X. 2008. Public expenditure, growth and poverty reduction in rural Uganda. *African Development Review* 20(3), 466–496.

Fan, S., Hazell, P. & Thorat, S. 2000. Government spending, agricultural growth and poverty in rural India. *American Journal of Agricultural Economics* 82(4), 1038–1051.

Fan, S., Zhang, L. & Zhang, X. 2004. Reforms, investment and poverty in rural China. *Economic Development and Cultural Change* 52(2), 395–421.

FAO. 1999. *Investment in Agriculture for Food Security: Situation and Resource Requirements to Reach the World Food Summit Objectives.* Committee on World Food Security. Rome.

FAO. 2001. *Mobilising Resources to Fight Hunger.* Committee on World Food Security. Rome.

FAO. 2002. *The World Food Summit: Five Years Later. Mobilizing the Political Will and Resources to Banish World Hunger.* Technical Background Documents. Rome.

FAO. 2009. *Global Agriculture Towards 2050.* Paper prepared for High level Expert Forum, Committee for Food Security. Rome.

FAO. 2011. *Estimates of Capital Stock in Agriculture (1975–2007).* Rome.

FAO. 2012a. *The State of Food and Agriculture 2012 – Investing in Agriculture for a Better Future.* Rome.

FAO. 2012b. *Level, Trend and Sources of Financing for Investment in Agriculture: A Review and Analysis of Available Sources of Data.* Rome.

FAO. 2012c. *Report of Workshop on Private Corporate Sector Investment in Agriculture in Southeast Asia.* Rome.

FAO. 2012d. *Proceedings of the Technical Workshop on Policies for Promoting Investment in Agriculture.* Rome.

FAO. 2013a. *Foreign Agriculture Investment Database.* (Available at: www.fao.org/investment/investment–policy/fdi)

FAO. 2013b. *Trends and Impacts of Foreign Investment in Developing Country Agriculture; Evidence from Case Studies.* Rome.

Fischer, R.A., Byerlee, D. & Edmeades, G.O. 2009. *Can Technology Deliver on the Yield Challenge to 2050?* Paper prepared for the Expert Meeting on How to Feed the World in 2050, organized by FAO, Rome, Italy, 24–26 June 2009.

Gandhi, V.P. 1990. *Investment Behaviour in Developing Countries: The Case of Agriculture in India.* IFRI studies. Vol.22, No.1. Washington, DC, IFPRI.

Gandhi, V.P. 1996. Investment behaviour in Indian agriculture. *Indian Journal of Agriculture Economics,* Vol.51, No.4.

Gonzalez-Velosa, C. 2011. *The Effects of Emigration and Remittances on Agriculture: Evidence from the Philippines.* Job market paper (available at http://econweb.umd.edu/~gonzalez-velosa/JMP_Gonzalezvelosa_JAN.pdf).

Grosh, M., Del Ninno, C., Tesliuc, E. & Ouerghi, A. 2008. *For Protection and Promotion: The Design and Implementation of Effective Safety Nets.* Washington, DC, World Bank.

Gulati, A. & Bathla, S. 2002. *Capital Formation in Indian Agriculture: Trends, Composition and Implications for Growth.* Occasional paper 24. Mumbai, India, National Bank for Agriculture and Rural Development.

Hang, S.C., Socheth, H. & Chandarany, O. 2011. *Foreign Investment in Agriculture in Cambodia.* Prepared for FAO. Rome.

Harrod, R.F. 1939. An essay in dynamic theory. *Economic Journal* 49(193), 14–33.

Hazell, P. & Haddad, L. 2001. *Agricultural Research and Poverty Reduction.* Food, Agriculture and the Environment Discussion Paper 34. Washington, DC, IFPRI.

High Level Committee on Estimation of Saving and Investment. 2009. *Report of the High Level Committee on Estimation of Saving and Investment.* New Delhi, Ministry of Statistics and Programme Implementation, Government of India.

Hossain, I. 2012. *A Rapid Assessment of Migration and Remittances and Their Impact on Food Security, Agriculture and Rural Development.* Report prepared for FAO. Rome.

Kazi, I. 2012. *Impact of International Migration and Remittances on Agriculture: The Case of Bangladesh.* Report prepared for FAO. Rome.

Krieckhaus, J. 2002. Reconceptualizing the developmental state: public savings and economic growth. *World Development* 30(10), 1697–1712.

Krueger, A., Schiff, M. & Valdes, A. 1988. Agricultural incentives in developing countries: measuring the effects of sectoral and economy wide policies. *World Bank Economic Review* 2(3), 255–272.

Krueger, A., Schiff, M. & Valdes, A. 1991. *The Political Economy of Agricultural Pricing Policy.* Baltimore, Johns Hopkins University Press for the World Bank.

Lewis, W.A. 1955. *The Theory of Economic Growth.* Homewood, Illinois, Richard D. Irwin.

Loayza, N. & Shankar, R. 2000. Private saving in India. *World Bank Economic Review* 14(3), 571–594.

Loayza, N., Schmidt-Hebbel, K. & Serven, L. 2000. *What Drives Private Saving Around the World?* Policy Research Working Paper 2309. Washington, DC, World Bank.

Lowder, S., Carisma, B. & Skoet, J. 2012. *Who Invests in Agriculture and How Much? An Empirical Review of The Relative Size of Various Investments in Agriculture in Low- and Middle-Income Countries.* ESA Working Paper No.12–09, Rome, FAO.

Lucas, R.E.B. 1987. Emigration to South Africa's mines. *American Economic Review* 77(3), 313–330.

Masson, P., Bayoumi, T. & Samiei, H. 1998. International evidence on the determinants of private saving. *World Bank Economic Review,* Vol. 12, pp. 483–501. Washington, DC, World Bank.

Mendola, M. 2008. Migration and technological change in rural households: complements or substitutes? *Journal of Development Economics* 85(1-2), 150–175.

Meyer, R. 2011. *Subsidies as an Instrument in Agriculture Finance: A Review.* Joint Discussion Paper of the Joint Donor CABFIN Initiative. Washington, DC, World Bank.

Migot-Adholla, S.E., Hazell, P.B.R., Blarel, B. & Place, F. 1991. Indigenous land rights systems in sub-Saharan Africa: a constraint on productivity? *World Bank Economic Review* 5(l), 155–175.

Mishra, S.N. 1996. Capital formation and accumulation in Indian agriculture since independence. *Indian Journal of Agricultural Economies* 51(1/2).

Mishra, S.N. & Chand, R. 1995. Public and private capital formation in indian agriculture: comments on the complementarity hypothesis and others. *Economic and Political Weekly* 30(25), A64–A79.

Mishra, S.N. & Hazell, P. 1996. Terms of trade, rural poverty, technology and investment. *Economic and Political Weekly* 31(13), A2–A13.

Misra, V.N. 1998. Economic reforms, terms of trade, aggregate supply and private investment in agriculture: Indian experience. *Economic and Political Weekly* 33(31), 2105–2109.

Mitra, A. 1996. Public and private investments in agriculture. In: B.M. Desai, ed. *Agricultural Development Paradigm for the Ninth Plan under New Economic Environment.* New Delhi, Oxford and IBH.

Mogues, T. 2011. The bang for the birr: public expenditures and rural welfare in Ethiopia. *Journal of Development Studies* 47(5), 735–752.

Mu, R. & van de Walle, D. 2007. *Rural Roads and Local Market Development in Vietnam.* Policy Research Working Paper 4340. Washington, DC, World Bank.

Muhleisen, M. 1997. *Improving India's Saving Performance.* IMF Working Paper No. 97/4. Washington, DC, IMF.

Oberai, A. & Singh, H.K.M. 1983. *Causes and Consequences of Internal Migration.* Delhi, Oxford University Press.

OECD. 2011. *Policy Framework for Investment in Agriculture: Policy Guidance for Promoting Private Investment in Agriculture in Africa.* Paris.

Olson, M. 1965. *The Logic of Collective Action.* New Haven, Yale University Press.

Ostrom, E. 1990. *Governing the Commons: The Evolution of Institutions for Collective Action.* Cambridge, UK, Cambridge University Press.

Paarlberg, R.L. & Grindle, M.S. 1991. Policy reform and policy myopia: agriculture in developing countries. *Food Policy* 16(5), 383–394.

Rao, C.H.H. 1997. *Agricultural Growth, Sustainability, and Poverty Alleviation in India.* Lecture presented at IFPRI, Washington, DC, IFPRI.

Rao, C.H.H. & Gulati, A. 1994. Indian agriculture: emerging perspectives and policy issues. *Economic and Political Weekly* 29(53), A158–A169.

Rath, N. 1989. Agricultural growth and investment in India. *Journal of Indian School of Political Economy,* Vol 1, No.1 January–June.

Reardon, T. & Barrett, C.B. 2000. Agroindustrialization, globalization, and international development: an overview of issues, patterns, and determinants. *Agricultural Economics* 23(3), 195–205.

Robinson, J. 1956. *The Accumulation of Capital.* London, Macmillan & Co. Ltd.

Rosenstein-Rodan, P.N. 1961. International aid for underdeveloped countries. *Review of Economics and Statistics* XLIII(2), 107–138.

Roy, B.C. 2001. *Investment and Productivity in Indian Agriculture.* PhD Thesis Submitted to Post Graduate School, Indian Agriculture Research Institute, New Delhi.

Sachs, J.D., McArthur, J.W., Schmidt-Traub, G., Kruk, M., Bahadur, C., Faye, M. & McCord, G. 2004. Ending Africa's poverty trap. *Brookings Papers on Economic Activity,* 1.

Salz, I.S. 1999. An examination of the causal relationship between savings and growth in the third world. *Journal of Economics and Finance* 23(1), 90–98.

Schmidhuber, J., Bruinsma, J. & Boedeker, G. 2009. *Capital Requirements for Agriculture in Developing Countries to 2050.* Paper prepared for the Expert Meeting on How to Feed the World in 2050. Rome.

Sen, B. 2003. Drivers of escape and descent: changing household fortunes in rural Bangladesh. *World Development* 31(3), 513–534.

Shetty, S.L. 1990. Investment in agriculture: brief review of recent trends. *Economic and Political Weekly* 25(7/8), 389–398.

Sinha, D. & Sinha, T. 1998. Cart before horse? The savings–growth nexus in Mexico. *Economics Letter* 61, 43–47.

Storm, S. 1993. *Macroeconomic Considerations in the Choice of an Agricultural Policy: a Study into Sectoral Independence with Reference to India.* Aldershot, UK, Avebury.

Suphannachart, W. & Warr, P. 2011. Research and productivity in Thai agriculture. *Australian Journal of Agricultural and Resource Economics* 55(1), 35–52.

Swinnen, J. & Maertens, M. 2007. Globalization, privatization and vertical coordination in food value chains in developing and transition countries. *Agricultural Economics* 37(s1), 89–102.

Todaro, M.P. & Smith, S.C. 2003. *Economic Development.* New York, Addison Wesley.

Tsegai, D. 2004. *Effects of Migration on the Source Communities in the Volta Basin of Ghana: Potential Links of Migration, Remittances, Farm and Non-Farm Self-Employment Activities.* Economics and Technological Change Working Paper. Bonn, University of Bonn.

UNCTAD. 2008. *Trade and Development Report 2008.* New York and Geneva, United Nations.

Uphoff, N. 2001. Balancing development and environmental goals through community-based natural resource management. In: D.R. Lee & C.B. Barrett, eds. *Tradeoffs or Synergies? Agricultural Intensification, Economic Development and the Environment.* Wallingford, UK, CAB International.

van der Mensbrugghe, D. 2005. *Linkage Technical Reference Document, Version 6.0.* Prepared by Development Prospects Group (DECPG). Washington, DC, World Bank.

von Cramon-Taubadel, S., Anriquez, G., de Haen, H. & Nivyevskiy, O. 2009. *Investment in Developing Countries' Food and Agriculture.* Paper prepared for the Expert Meeting on How to Feed the World in 2050. Rome.

Vorley, B., Lundy, M. & MacGregor, J. 2009. Business models that are inclusive of small farmers. In: C. da Silva, D. Baker, A.W. Shepherd, G. Jenane & S. Miranda-da-Cruz, eds. *Agro-industries for Development,* pp. 186–222. Rome, FAO and UNIDO.

Waleerat, S. & Nipawan, T. 2011. *Analysis of International Investments in the Agricultural Sector of Thailand.* Prepared for FAO. Rome.

Warman, F. & Thirlwall, A.P. 1994. Interest rates, savings, investment and growth in mexico 1960–90: tests of the financial liberalisation hypothesis. *Journal of Development Studies* 30(3), 629–649.

Wilcox, W.W. 1943. Capital in agriculture. *Quarterly Journal of Economics* 58(1), 49–64.

World Bank. 2004. *World Development Report 2005.* Washington, DC, World Bank.

World Bank. 2010. *Project Appraisal Document, Report No: 55581-KH.* Washington, DC, World Bank.

World Bank. 2011. *Practitioners Toolkit or Agriculture Public Expenditure Analysis.* Washington, DC, World Bank and UK Department for International Development.

Index

Page numbers in **bold** refer to illustrations and tables

farmers (*continued*)

    incomes, investment relationship **46**, 47

    investments 5, 55–56, 61

    organizations 53, 54

    positive savings generation 64

    training and support need 54

farms

    family 42

    household investment 37

    household savings promoting 56–61

    non-viable, support programmes 66

    size 13, 48, 52, 64–65

    small 13, 41, 48, 52, 53, 54, 64–65, 70

    structures 17

    see also small- and medium-sized enterprises

fertilizers 5, 6, 12–13

finance sources 25–29, 55, 61, **73**

financial sector 51, 58–60, 68, 74

fishing, FDI inflows 30, **31**

food

    FDI inflows **32**, **34**

    security 60, 70, 72

    security crisis 1–2

foreign direct investment (FDI) 3, 30–34, 69, 71

forestry, FDI inflows 30, **31**

fragmentation, social 70

fresh fruit trade 50

gender discrimination 48

governance 58

government 5, 23, 68–69

    see also public sector

gross domestic product (GDP) **20**, 22, 26, 27, 28, 43, 61

gross domestic savings (GDS) 26, 28, 61

gross fixed capital formation (GFCF) 5

growth

    agricultural 2

    agricultural capital stock **37**

    rate 17, 20, 21, **22**, 28

    small- and medium-sized enterprises (SMEs) 29

    strategies 57, 64

guidance, international 72

hand tools, value 17

Harrod and Domar growth formula 4

High Level Committee on Estimation of Saving and Investment, India 56

Household Income and Expenditure Survey (2010), Bangladesh 61

human capital 11, 42, 48

    see also labour; workers

human resource development 74

hunger 1, 3, 19

    see also undernourishment

hunting, FDI inflows 30, **31**

ideas 62

ideology 62, 63

incentives 38–54

    see also subsidies

income 45–47, 60–61

India, macro-economic story 61

indicators 17, 18, **20**, **27**

industrialization 13, 62

    see also agro-industry

inflows 30, **31**, **32**, **33**

information systems, dependable 68

infrastructure 12–13, 17, 49, 67, 74

inputs 5, 6, 11, 13–14, 41

institutions 59, 67, 72

interest group approaches 62

interest rates impact 44

International Food Policy Research Institute (IFPRI) 28–29

interventions 64

interviews, private sector investors 29–30

inventories 5, 16, **17**, **19**, 54, **75**

investment promotion agencies (IPAs) 69

investment types 4

investors 23–34, 55

    see also farmers; foreign direct investment; private sector; public sector

Joint Statement on Global Food Security 2

labour 8–9, **10**, 11, 18, 19, 50

    see also human capital; workers